Junior Skill Builders

ALGEBRA
in
15 Minutes a Day

Junior Skill Builders

ALGEBRA
in
15 Minutes a Day

LEARNINGEXPRESS®

NEW YORK

Copyright © 2009 LearningExpress, LLC.

All rights reserved under International and Pan-American Copyright Conventions.
Published in the United States by LearningExpress, LLC, New York.

Library of Congress Cataloging-in-Publication Data:
Junior skill builders: algebra in 15 minutes a day.
    p. cm.
  ISBN 978-1-57685-673-4
  1. Algebra—Study and teaching (Middle school) 2. Algebra—Study and teaching (Secondary) I. LearningExpress (Organization) II. Title: Algebra in fifteen minutes a day.
  QA159.A443 2009
  512.0071'2—dc22

                                                2008042665

Printed in the United States of America

10 9 8 7 6 5 4 3 2 1

For more information or to place an order, contact LearningExpress at:
  2 Rector Street
  26th Floor
  New York, NY 10006

Or visit us at:
  www.learnatest.com

C O N T E N T S

Junior Skill Builders

ALGEBRA
in
15 Minutes a Day

INTRODUCTION

IF YOU'VE NEVER seen algebra before, then the math you have seen has mostly involved numbers. You may know that algebra seems to involve the letter x a whole lot, and that may be enough to make you want to close this book. But really, algebra is just about finding missing numbers. We do not always know the value of a quantity, so we use a letter like x to hold the place of that value.

Algebra can help us find how much interest our money is earning in a bank, or algebra might help us find the probability of drawing the ace of spades from a deck of cards. You might think that you do not use algebra often in everyday life, but any time you ask yourself a question like, "How many magazines can I buy with the money I have?" you are using algebra. The number of magazines you can buy is a number you do not know, and that number multiplied by the cost of one magazine must be less than or equal to the amount of money you have.

USING THIS BOOK

This book is divided into 30 lessons, plus a pretest and a posttest. The lessons are designed to be completed in 15 minutes. Each day, you can learn a little algebra without taking too much of your time.

THE BOOK AT A GLANCE

Each lesson in this book builds on the previous lesson. We start with an explanation of algebra and the words that are often used when we are learning algebra. What is a term? What is a variable? What is a constant? If you already know the answers to these questions, you are off to a fast start. If not, do not worry; that is what this book is for—to introduce you to algebra and show you how to use it to solve a variety of problems.

Before the lessons begin, we'll start with a pretest. This 30-question test will give you an idea of how well you know algebra now. Every question contains an answer explanation and lets you know from which lesson in the book the question came. Remember, it is only a pretest—if you knew the answer to every question, you would not be reading this book!

After you have been working through all 30 lessons, the book ends with a posttest. The posttest, like the pretest, is 30 questions. Compare your results on the posttest to your pretest. Did you correctly answer the types of questions you struggled with in the pretest? Are there a few topics that you might need to spend more time reviewing? You can always go back and reread a lesson or two. Each lesson contains plenty of practice problems, with complete answer explanations for every question, plus tips to help you remember how to solve a problem or another way to go about solving problems.

The pretest is next—grab a pen or pencil and get started!

THIS PRETEST WILL test your knowledge of basic algebra. The 30 questions are presented in the same order that the topics they cover are presented in this book. The book builds skills from lesson to lesson, so skills you have learned in earlier lessons may be required to solve problems found in later lessons. This means that you may find the problems toward the beginning of the test easier to solve than the problems toward the end of the test.

For each of the 30 multiple-choice questions, circle the correct answer. The pretest will show you which algebra topics are your strengths and which topics you might need to review—or learn for the first time. Explanations are provided for every question, along with the lesson number that teaches the skills needed to answer the question. After taking the pretest, you might realize that you need only to review a few lessons. Or you might benefit from working through the book from start to finish.

Save your answers after you have completed the pretest. When you have finished the lessons in this book, take the posttest. Compare your pretest to your posttest to see how you have improved.

PRETEST

1. $12q^6 + 4q^6 =$
   a. $8q^6$
   b. 16
   c. $16q^6$
   d. $16q^{12}$

2. $(-45a^4b^9c^5) \div (9ab^3c^3) =$
   a. $-5a^3b^6c^2$
   b. $-5a^4b^3c^3$
   c. $-5a^4b^3c^2$
   d. $-5a^5b^{12}c^8$

3. What is $10 + 6j$ when $j = -3$?
   a. -18
   b. -8
   c. 8
   d. 28

4. Simplify $-7g^6 + 9h + 2h - 8g^6$.
   a. $-4g^6h$
   b. $-2g^6 - 4h$
   c. $-5g^6 + h$
   d. $-15g^6 + 11h$

5. Which of the following is nine less than three times a number?
   a. $x - 9$
   b. $9 - 3x$
   c. $3x - 9$
   d. $3(x - 9)$

6. Which of the following is equal to $12x^7 - 24x^3$?
   a. $-12x^4$
   b. $12x^3(x^4 - 2)$
   c. $12x(x^7 - 2x^3)$
   d. $12x(x^6 - 2x)$

7. $(4w^9)^3 =$
   a. $4w^{12}$
   b. $4w^{27}$
   c. $12w^{27}$
   d. $64w^{27}$

8. What is the value of z in $z - 7 = -9$?
   a. -2
   b. -1
   c. 2
   d. 16

9. If $4n = -28m$, what is n in terms of m?
   a. $-32m$
   b. $-22m$
   c. $-7m$
   d. $7m$

10. If $-7k - 11 = 10$, what is the value of k?
   a. -3
   b. -1
   c. 2
   d. 21

11. What is the value of p in $\frac{p}{6} + 13 = p - 2$?
   a. 6
   b. 12
   c. 15
   d. 18

12. Which of the following is equivalent to $\sqrt[3]{o^{27}}$?
   a. o^3
   b. o^9
   c. o^{24}
   d. o^{81}

13. Find the slope of the equation $8y = 16x - 4$.

   a. $-\frac{1}{2}$

   b. 2

   c. 8

   d. 16

14. What equation was used to build the input/output table below?

x	y
1	7
2	10
3	13
4	16
5	19

   a. $y = 3x + 4$

   b. $y = 4x - 1$

   c. $y = 5x - 2$

   d. $y = 7x$

15. Which of the following is the graph of $y = -4x - 3$?

   a.

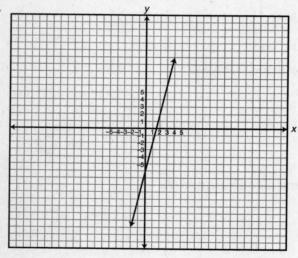

b.

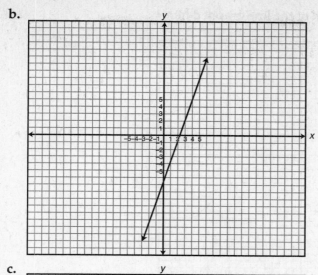

c.

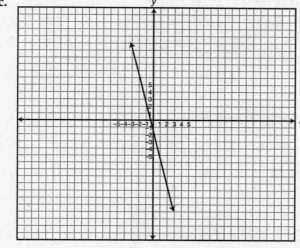

d.

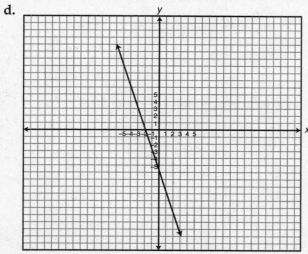

16. What is the equation of the line graphed here?

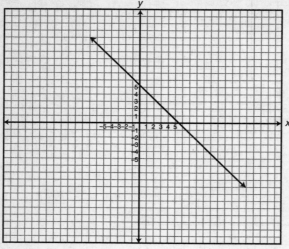

a. $y = x + 7$
b. $y = x - 7$
c. $y = -x - 7$
d. $y = -x + 7$

17. Find the distance from (−3,−3) to (9,13).

a. 12 units
b. 16 units
c. 20 units
d. 24 units

18. Which of the following is NOT a function?

a. $y = x^7$
b. $x = y^2$
c. $x = \sqrt{y}$
d. $y = |x|$

19. What is the solution to the following system of equations?

$4y + x = 7$
$3y - 2x = 30$

a. $x = -9, y = 4$
b. $x = 3, y = 1$
c. $x = 3, y = 12$
d. $x = -6, y = 6$

20. What inequality is represented by the following graph?

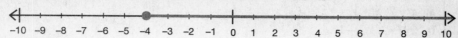

   a. $x < -4$

   b. $x > -4$

   c. $x \leq -4$

   d. $x \geq -4$

21. The solution set of what system of inequalities is shown as the darker region on the following graph?

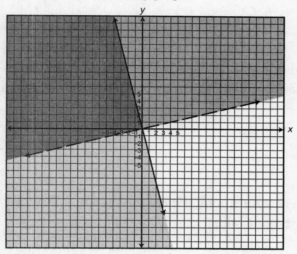

   a. $y > \frac{1}{4}x,\ y \leq -4x$

   b. $y < \frac{1}{4}x,\ y \leq -4x$

   c. $y > \frac{1}{4}x,\ y \geq -4x$

   d. $y < \frac{1}{4}x,\ y \geq -4x$

22. $(7x - 1)(x + 2) =$

   a. $7x^2 - 2$

   b. $7x^2 + 13x - 2$

   c. $7x^2 + 14x - 2$

   d. $7x^2 - 14x - 2$

23. Which of the following shows $x^2 + 10x + 24$ correctly factored?

   a. $(x + 5)(x + 5)$

   b. $(x + 3)(x + 8)$

   c. $(x + 12)(x + 2)$

   d. $(x + 4)(x + 6)$

24. Find the roots of $x^2 + 6x = 7$.

   a. $x = 7, x = 1$
   b. $x = -7, x = -1$
   c. $x = -7, x = 1$
   d. $x = 7, x = -1$

25. Lindsay's bowling score is seven less than three times Jamie's score. If Lindsay's score is 134, what is Jamie's score?

   a. 39
   b. 41
   c. 43
   d. 47

26. The ratio of yellow beads to green beads on a necklace is 3:10. If there are 60 yellow beads on the necklace, how many green beads are on the necklace?

   a. 70
   b. 100
   c. 200
   d. 600

27. The mean of a set of seven values is 15. If six of the values are 9, 19, 17, 16, 11, and 20, what is the seventh value?

   a. 11
   b. 13
   c. 15
   d. 16

28. What is 48 after a 55% increase?

   a. 26.4
   b. 30.9
   c. 69.6
   d. 74.4

29. What is the value of the ninth term of the following arithmetic sequence?
$-3x + 1, x - 7, 2x - 3, 3x + 1, \ldots$

   a. 53
   b. 61
   c. 64
   d. 75

30. The width of a rectangle is four times its length. If the perimeter of the rectangle is 560 meters, what is the width of the rectangle?

  a. 56 m

  b. 112 m

  c. 224 m

  d. 448 m

ANSWERS

1. c. Each term has a base of q and an exponent of 6, so the base and exponent of your answer is q^6.

  Add the coefficients: $12 + 4 = 16$, $12q^6 + 4q^6 = 16q^6$.

  For more on this skill, review Lesson 2.

2. a. Divide the coefficients: $-45 \div 9 = -5$. Carry the bases (a, b, and c) from the dividend into the answer. There are no bases in the divisor that are not in the dividend.

  Subtract the exponent of a in the divisor from the exponent of a in the dividend: $4 - 1 = 3$. The exponent of a in the answer is 3.

  Subtract the exponent of b in the divisor from the exponent of b in the dividend: $9 - 3 = 6$. The exponent of b in the answer is 6.

  Subtract the exponent of c in the divisor from the exponent of c in the dividend: $5 - 3 = 2$. The exponent of c in the answer is 2.

$$(-45a^4b^9c^5) \div (9a^2b^3c^3) = -5a^3b^6c^2$$

  For more on this skill, review Lesson 3.

3. b. Replace j with -3:

$$10 + 6(-3)$$

  Multiply before adding:

$$6(-3) = -18$$

  The expression becomes $10 + -18$.

$$\text{Add: } 10 + -18 = -8.$$

  For more on this skill, review Lesson 4.

4. d. This expression has two g^6 terms and two h terms.

  Combine the g^6 terms: $-7g^6 - 8g^6 = -15g^6$.

  Combine the h terms: $9h + 2h = 11h$.

  $-7g^6 + 9h + 2h - 8g^6$ simplifies to $-15g^6 + 11h$.

  For more on this skill, review Lesson 5.

5. c. The keyword phrase *less than* signals subtraction and the keyword *times* signals multiplication. Three times a number is $3x$. Nine less than that quantity is $3x - 9$.

For more on this skill, review Lesson 6.

6. b. These terms are unlike, so they cannot be combined, but they can be factored.

The factors of $12x^7$ are 1, 2, 3, 4, 6, 12, and x^7, and the factors of $-24x^3$ are 1, 2, 3, 4, 6, 8, 12, 24, and x^3 (and their negatives).

Both terms have 12 and the variable x as common factors. The smaller exponent of x between the two terms is 3, so we can factor $12x^3$ out of both terms.

Divide both terms by $12x^3$: $12x^7 \div 12x^3 = x^4$, and $-24x^3 \div 12x^3 = -2$. $12x^7 - 24x^3$ factors into $12x^3(x^4 - 2)$.

For more on this skill, review Lesson 7.

7. d. $4w^9$ is raised to the third power.

Raise 4 to the third power and raise w^9 to the third power. $4^3 = 64$.

To find $(w^9)^3$, multiply the exponents: $(9)(3) = 27$, $(w^9)^3 = w^{27}$, and $(4w^9)^3 = 64w^{27}$.

For more on this skill, review Lesson 8.

8. a. In the equation $z - 7 = -9$, 7 is subtracted from z. Use the opposite operation, addition, to solve the equation. Add 7 to both sides of the equation:

$$z - 7 + 7 = -9 + 7$$
$$z = -2$$

For more on this skill, review Lesson 9.

9. c. To find n in terms of m, we must get n alone on one side of the equation, with m on the other side of the equals sign. In the equation $4n = -28m$, n is multiplied by 4. Use the opposite operation, division, to get n alone on one side of the equation. Divide both sides of the equation by 4:

$$\frac{4n}{4} = \frac{-28m}{4}$$
$$n = -7m$$

The value of n, in terms of m, is $-7m$.

For more on this skill, review Lesson 9.

10. a. The equation $-7k - 11 = 10$ shows multiplication and subtraction. We will need to use their opposites, division and addition, to find the value of k. Add first:

$$-7k - 11 + 11 = 10 + 11$$
$$-7k = 21$$

Because k is multiplied by -7, divide both sides of the equation by -7:

$$\frac{-7k}{-7} = \frac{21}{-7}$$
$$k = -3$$

For more on this skill, review Lesson 10.

11. d. The equation $\frac{p}{6} + 13 = p - 2$ shows division, addition, and subtraction. The variable p appears on both sides of the equation, and a constant is on both sides of the equation. Start by subtracting $\frac{p}{6}$ from both sides. This will leave us with just one p term:

$$\frac{p}{6} - \frac{p}{6} + 13 = p - \frac{p}{6} - 2$$
$$13 = \frac{5}{6}p - 2$$

Because 2 is subtracted from $\frac{5}{6}p$, add 2 to both sides of the equation:

$$13 + 2 = \frac{5}{6}p - 2 + 2$$
$$15 = \frac{5}{6}p$$

Finally, because p is multiplied by $\frac{5}{6}$, multiply both sides of the equation by $\frac{6}{5}$:

$$\frac{6}{5}(15) = \frac{6}{5}\left(\frac{5}{6}p\right)$$
$$18 = p$$

For more on this skill, review Lesson 10.

12. b. The third root of o^{27} is the quantity that, when multiplied three times, is equal to o^{27}. To find the third root of a base, divide the exponent of the base by 3: $27 \div 3 = 9$. $(o^9)(o^9)(o^9) = o^{27}$, which is why $\sqrt[3]{o^{27}} = o^9$.

For more on this skill, review Lesson 11.

13. b. The equation $8y = 16x - 4$ is not in slope-intercept form, because y is not alone on one side of the equation. Since y is multiplied by 8, we must divide both sides of the equation by 8:

$$\frac{8y}{8} = \frac{16x - 4}{8}$$
$$y = 2x - \frac{1}{2}$$

The equation is now in slope-intercept form. The slope is the coefficient of x, so the slope of this line is 2.

For more on this skill, review Lesson 12.

14. a. First, find the slope. Take the difference between the first two y values in the table and divide it by the difference between the first two x values in the table: $\frac{10-7}{2-1} = \frac{3}{1} = 3$. The slope of the line is 3. Next, find the y-intercept using the equation $y = mx + b$. Substitute 3, the slope of the line, for m. Substitute the values of x and y from the first row of the table, and solve for b, the y-intercept:

$$7 = 3(1) + b$$
$$7 = 3 + b$$
$$4 = b$$

The slope of the line is 3 and the y-intercept is 4. The equation of this line is $y = 3x + 4$.

For more on this skill, review Lesson 13.

15. c. The slope of the graph $y = -4x - 3$ is -4, since the equation is in slope-intercept form and the coefficient of x is -4. The y-intercept of the equation is -3. The graphs in choices **a** and **c** have y-intercepts of -3, but only the graph in choice **c** has a slope of -4.

For more on this skill, review Lesson 14.

16. d. To find the equation of the line, begin by finding the slope using any two points on the line. When $x = 1$, $y = 6$, and when $x = 2$, $y = 5$. The slope is equal to: $\frac{5-6}{2-1} = -\frac{1}{1} = -1$. The y-intercept can be found right on the graph. The line crosses the y-axis where $y = 7$, which means that 7 is the y-intercept. This is the graph of the equation $y = -x + 7$.

For more on this skill, review Lesson 15.

17. c. Use the distance formula to find the distance between $(-3,-3)$ and $(9,13)$. Because $(-3,-3)$ is the first point, x_1 will be -3 and y_1 will be -3. $(9,13)$ is the second point, so x_2 will be 9 and y_2 will be 13:

$$D = \sqrt{(x_2 - x_1)^2 + (y_2 - y_1)^2}$$
$$D = \sqrt{(9 - (-3))^2 + (13 - (-3))^2}$$
$$D = \sqrt{(12)^2 + (16)^2}$$
$$D = \sqrt{144 + 256}$$
$$D = \sqrt{400}$$
$$D = 20 \text{ units}$$

For more on this skill, review Lesson 16.

18. b. An equation is a function if every x value has no more than one y value. In the equation $x = y^2$, every positive x value will have two y values, since the square of a positive number and the square of its negative are the same. The equation $x = y^2$ is not a function.
For more on this skill, review Lesson 17.

19. a. This system could be solved using either substitution or elimination. Because the first equation can easily be solved for x, use substitution to solve.
Write x in terms of y by subtracting $4y$ from both sides of the first equation:

$$4y + x = 7$$
$$x = 7 - 4y$$

Replace x in the second equation with the expression that is equal to x, $7 - 4y$:

$$3y - 2(7 - 4y) = 30$$

Solve for the value of y:

$$3y - 2(7 - 4y) = 30$$
$$3y - 14 + 8y = 30$$
$$11y - 14 = 30$$
$$11y = 44$$
$$y = 4$$

Replace y with its value in either equation and solve for the value of x:

$$4(4) + x = 7$$
$$16 + x = 7$$
$$x = -9$$

For more on this skill, review Lesson 18 and Lesson 19.

20. d. The graph shows the number –4 circled, and the circle is solid. This means that –4 is part of the solution set. All of the values that are greater than –4 are highlighted, which means that the solution set is all values that are greater than or equal to –4: $x \geq -4$.
For more on this skill, review Lesson 20.

21. a. The solution set of two inequalities is the area of a graph where the solution set of one inequality overlaps the solution set of the other inequality.

The line $y = \frac{1}{4}x$ is dashed, which means that the points on the line are not part of the solution set.

The line $y = -4x$ is solid, which means that the points on the line are part of the solution set.

The point $(-4, 4)$ is in the solution set, so it can be used to test each inequality.

$\quad \frac{1}{4}(-4) = -1$, which is less than 4. Therefore, $y > \frac{1}{4}x$.

$\quad -4(-4) = 16$, which is greater than 4. Therefore, $y \leq -4x$.

This graph shows the solution set of $y > \frac{1}{4}x$ or $y \leq -4x$.

For more on this skill, review Lesson 21.

22. b. To find the product of two binomials, use FOIL and combine like terms:

$\quad$ First: $(7x)(x) = 7x^2$

$\quad$ Outside: $(7x)(2) = 14x$

$\quad$ Inside: $(-1)(x) = -x$

$\quad$ Last: $(-1)(2) = -2$

$\quad 7x^2 + 14x - x - 2 = 7x^2 + 13x - 2$

For more on this skill, review Lesson 22.

23. d. To factor $x^2 + 10x + 24$, begin by listing the positive and negative factors of the first and last terms:

Factors of x^2: $1, -x, x, x^2$

Factors of 24: $-24, -12, -8, -6, -4, -3, -2, -1, 1, 2, 3, 4, 6, 8, 12, 24$

x^2 is the square of either x or $-x$. Begin by trying x as the first term in each binomial:

$\quad (x + \underline{\quad})(x + \underline{\quad})$

The coefficients of each x term are 1. The last terms of each binomial must multiply to 24 and add to 10, since $10x$ is the sum of the Outside and Inside products. $10x$ is positive, so we are looking for two positive numbers that multiply to 24 and add to 10.

$\quad (2)(12) = 24$, but $2 + 12 = 14$

$\quad (3)(8) = 24$, but $3 + 8 = 11$

$\quad (4)(6) = 24$, and $4 + 6 = 10$

The constant of one binomial is 4 and the constant of the other binomial is 6:

$\quad (x + 4)(x + 6)$

Check the answer using FOIL:

First: $(x)(x) = x^2$

Outside: $(x)(6) = 6x$

Inside: $(4)(x) = 4x$

Last: $(4)(6) = 24$

$x^2 + 6x + 4x + 24 = x^2 + 10x + 24$

For more on this skill, review Lesson 23.

24. c. Before we can find the roots of $x^2 + 6x = 7$, it must be in the form $ax^2 + bx + c = 0$. Subtract 7 from both sides of the equation. The equation is now $x^2 + 6x - 7 = 0$.

Factor $x^2 + 6x - 7$ and set each factor equal to 0.

Factors of x^2: 1, $-x$, x, x^2

Factors of -7: -7, -1, 1, 7

Because the last term is -7, we are looking for a negative number and a positive number that multiply to -7:

$(x - 7)(x + 1) = x^2 - 6x - 7$. The middle term is too small.

$(x + 7)(x - 1) = x^2 + 6x - 7$. These are the correct factors.

Set $x + 7$ and $x - 1$ equal to 0 and solve for x:

$x + 7 = 0$ $x - 1 = 0$

$x = -7$ $x = 1$

The roots of $x^2 + 6x - 7 = 0$ are -7 and 1.

For more on this skill, review Lesson 24.

25. d. We are looking for Jamie's score, so we can use x to represent that number. Lindsay's score is seven less than three times that number, which means that her score is $3x - 7$. We know that her score is 134, so we can set $3x - 7$ equal to 134 and solve for x:

$3x - 7 = 134$

$3x = 141$

$x = 47$

Jamie's score is 47.

For more on this skill, review Lesson 25.

26. c. Write the ratio of yellow beads to green beads as a fraction. $3:10 = \frac{3}{10}$. The number of green beads is unknown, so represent that number with x. The ratio of actual yellow beads to actual green beads is $60:x$, or $\frac{60}{x}$. Set these ratios equal to each other, cross multiply, and solve for x:

$\frac{3}{10} = \frac{60}{x}$

$3x = 600$

$x = 200$

There are 200 green beads on the necklace.

For more on this skill, review Lesson 26.

27. b. The mean of a set is equal to the sum of the values divided by the number of values. Let x represent the seventh value in the set:

$$9 + 19 + 17 + 16 + 11 + 20 + x = 92 + x$$
$$\frac{92 + x}{7} = 15$$
$$92 + x = 105$$
$$x = 13$$

The seventh value of the set is 13.

For more on this skill, review Lesson 27.

28. d. The original value is 48 and the new value is x. Percent increase is equal to the new value minus the original value divided by the original value.

Subtract 48 from x and divide by 48. Set that fraction equal to 55%, which is 0.55:

$$\frac{x - 48}{48}$$

Multiply both sides of the equation by 48, and then add 48 to both sides:

$$x - 48 = 26.4$$
$$x = 74.4$$

48 after a 55% increase is 74.4.

For more on this skill, review Lesson 28.

29. a. The formula $t_n = t_1 + (n - 1)d$ gives us the value of any term in an arithmetic sequence. The first term, t_1, is $-3x + 1$.

The difference, d, can be found by subtracting any pair of consecutive terms: $x - 7 - (-3x + 1) = 4x - 8$. $d = 4x - 8$. Because we're looking for the ninth term, $n = 9$:

$$t_9 = -3x + 1 + (9 - 1)(4x - 8)$$
$$t_9 = -3x + 1 + (8)(4x - 8)$$
$$t_9 = -3x + 1 + 32x - 64$$
$$t_9 = 29x - 63$$

The ninth term in the sequence is $29x - 63$.

The difference between any two consecutive terms in an arithmetic sequence is always the same. The difference between the first term and the second term is $4x - 8$. The difference between the second term and the third term is $2x - 3 - (x - 7) = x + 4$.

Set these two differences equal to each other and solve for x:

$$4x - 8 = x + 4$$
$$3x - 8 = 4$$
$$3x = 12$$
$$x = 4$$

The value of x in this sequence is 4, which means that the ninth term, $29x - 63$, is equal to $29(4) - 63 = 116 - 63 = 53$.

For more on this skill, review Lesson 29.

30. c. The formula for perimeter of a rectangle is $P = 2l + 2w$.

Let x represent the length of the rectangle. The width is 4 times the length, which means that it is $4x$. Substitute 560 for P, x for l, and $4x$ for w:

$$560 = 2(x) + 2(4x)$$
$$560 = 2x + 8x$$
$$560 = 10x$$
$$56 = x$$

Because $x = 56$, the length of the rectangle is 56 m. The width is four times the length, which means that the width is $4(56) = 224$ m.

For more on this skill, review Lesson 30.

ⓢ ⓔ ⓒ ⓣ ⓘ ⓞ ⓝ 1

algebra basics

BEFORE WE CAN use algebra, we need to understand what it is. This section begins by explaining the vocabulary of algebra, so that when you see x in a problem, you will know what it is and why it's there. Once these definitions are out of the way, we will review how to perform basic operations (addition, subtraction, multiplication, and division) on real numbers, and then show how these same operations can be performed on algebraic quantities.

Just as we can write number sentences that add or subtract two numbers, we can write sentences, or *expressions*, using algebra. These expressions might be given to us in words, so we will learn how to turn these words into expressions that contain numbers and variables. Then, we will review factoring and exponents, and show how these topics relate to algebra, too.

This section reviews some basic math skills and introduces you to the basics of algebra, including:

- the language of algebra
- signed numbers
- like and unlike terms
- order of operations
- single-variable and multivariable expressions
- factoring
- positive, negative, and fractional exponents

explaining variables and terms

The human mind has never invented a
labor-saving machine equal to algebra.

—AUTHOR UNKNOWN

In this lesson, you'll learn the language of algebra, how to define variables and terms, and a short review of integers.

MATH TOPICS ALWAYS seem to have scary sounding names: trigonometry, combinatorics, calculus, Euclidean plane geometry—and algebra. What is algebra? **Algebra** is the representation of quantities and relationships using symbols. More simply, algebra uses letters to hold the place of numbers. That does not sound so bad. Why do we use these letters? Why not just use numbers? Because in some situations, we do not always have all the numbers we need.

Let's say you have 2 apples and you buy 3 more. You now have 5 apples, and we can show that addition by writing the sentence $2 + 3 = 5$. All of the values in the sentence are numbers, so it is easy to see how you went from 2 apples to 5 apples.

Now, let's say you have a beaker filled with 134 milliliters of water. After pouring more water into the beaker, you look closely and see that you now have 212 milliliters of water. How much water was added to the beaker? Before you perform any mathematical operation, that quantity of water is unknown.

TIP: If we do not know the value of a quantity in a problem, that value is an **unknown**.

We can write a sentence to show what happened to the volume of water in the beaker even though we don't know how much water was added. A symbol can hold the place of the quantity of water that was added. Although we could use any symbol to represent this quantity, we usually use letters, and the most commonly used letter in algebra is x.

There is no clear reason why x came to be used most often to represent unknowns. René Descartes, a French mathematician, was one of the first to use x, y, and z to represent unknown quantities—back in 1637! While many have tried to determine why he used these letters, no one knows for certain.

The beaker had 134 milliliters of water in it when x milliliters were added to it. Read that sentence again. We describe the unknown quantity in the same way we would a real number. When a symbol, such as x, takes the place of a number, it is called a **variable**. We can perform the same operations on variables that we perform on real numbers. After x milliliters are added to the beaker, the beaker contains 212 milliliters. We can write this addition sentence as $134 + x = 212$. Later in this book, we will learn how to solve for the value of x and other variables.

In the sentence $134 + x = 212$, 134 and 212 are numbers and x is a variable. Because the variable x holds the place of a number, we can perform the same operations on it that we would perform on a number.

We can add 4 to the variable x by writing $x + 4$. We can subtract 4 from x by writing $x - 4$. Multiplication we show a little bit differently. Because the letter x looks like the multiplication symbol ($\times$), we show multiplication by placing the number that multiplies the variable right next to the variable, with no space. To show 4 multiplied by x, we write $4x$. There is no operation symbol between 4 and x, and that tells us to multiply 4 and x. Multiplication is sometimes shown by two adjacent sets of parentheses. Another way to show 4 multiplied by x is $(4)(x)$.

Division is most often written as a fraction. x divided by 4 is $\frac{x}{4}$. This could also be written as $\frac{x}{4}$ or $x \div 4$, but these notations are less common.

ALGEBRA VOCABULARY

A sentence, whether it contains variables or not, is made up of terms. A **term** is a variable, constant, or product of both, with or without exponents, and is usually separated from another term by addition, subtraction, or an equal sign.

While a variable can have different values in different situations, a **constant** is a term that never changes value. Real numbers are constants. The sentence $x + 4 = 5$ contains 3 terms: x, 4, and 5. x is a variable, and 4 and 5 are constants.

The sentence $3x - 5 = 11$ also contains 3 terms: $3x$, -5, and 11. $3x$ is a single term, because 3 and x are multiplied, not added or subtracted. In the same way, $\frac{x}{7} - 1 = 2$ also has only three terms, because $\frac{x}{7}$ is a single term.

PRACTICE 1

For each problem, write the number of terms and list the variables.

1. $x - 6$

2. $a + b$

3. $3f + g$

4. $5p + 6q - 7r$

5. $xy + 2z$

The numerical multiplier, or factor, of a term is called a **coefficient**. In the term $3x$, 3 is the coefficient of x, because 3 multiplies x. In the term $9y$, the coefficient of y is 9. In multiplication, the order in which one value multiplies another does not matter: 4×5 and 5×4 both equal 20. The order in which 3 and x are multiplied does not matter, either, but we typically place the constant in front of the variable. The constant is considered the coefficient, and the variable is considered the **base**. Because the coefficient is one factor of the term, the base is the other factor. If a variable appears to have no coefficient, then it has a coefficient of 1: $1x = x$.

....

TIP: Division can be rewritten as multiplication. x divided by 5 is the same as x multiplied by $\frac{1}{5}$. The coefficient of x in the term $\frac{x}{5}$ is $\frac{1}{5}$, because $\frac{x}{5}$ can be written as $\frac{1}{5}x$.

....

In algebra, the base of a term is often raised to an **exponent**. An exponent is a constant or variable that tells you how many times a base must be multiplied by itself. Exponents are small numbers (superscripts) that appear above and to the right of a base. The term x^2 is equal to x multiplied by x. The term y^6 means $(y)(y)(y)(y)(y)(y)$. If a variable appears to have no exponent, then it has an exponent of 1: $x^1 = x$.

PRACTICE 2

For each term, find the coefficient, the base, and the exponent.

1. $5x^3$

2. z^5

3. $\frac{b}{3}$

4. $8m$

5. $\frac{2}{9}c^7$

LIKE AND UNLIKE TERMS

If two terms have the same base raised to the same exponent, then the two terms are called **like terms**. For instance, $2a^2$ and $-6a^2$ are like terms, because both have a base of a with an exponent of 2. Even though the terms have different coefficients, they are still like terms. If two terms have different bases, or identical bases raised to different exponents, then the two terms are **unlike terms**. $7m$ and $7n$ are unlike because they have different bases. $7m^4$ and $7m^5$ are also unlike terms. Even though they have the same base, the exponents of the bases are different. In the next lesson, we will see why recognizing terms as like or unlike is so important.

PRACTICE 3

Label each pair of terms as "like" or "unlike."

1. $3p$ and $3q$

2. $-k^6$ and $12k^6$

3. $\frac{3}{8}c^2$ and $8c^2$

4. $8m^4$ and $8n^4$

5. $5v^{10}$ and $3v^{-10}$

ANSWERS

Practice 1

1. There are two terms, because x and 6 are separated by subtraction; x is the only variable.
2. There are two terms, because a and b are separated by addition; a and b are both letters, so they are both variables.
3. There are two terms, because $3f$ and g are separated by addition; f and g are both letters, so they are both variables.
4. There are three terms, because $5p$ and $6q$ are separated by addition and $6q$ and $7r$ are separated by subtraction; p, q, and r are all variables.
5. There are two terms, because xy and $2z$ are separated by addition; x, y, and z are all variables.

Practice 2

1. In the term $5x^3$, the number 5 multiplies x^3, so 5 is the coefficient and x is the base. The exponent of the base is 3.
2. The term z^5 has no number in front of z^5, so the coefficient of the term is 1. The base is z and the exponent is 5.
3. The term $\frac{b}{3}$ can be rewritten as $\frac{1}{3b}$. Since b is multiplied by $\frac{1}{3}$, the coefficient of the term is $\frac{1}{3}$ and the base is b. b has no small number written above it and to the right, so it has an exponent of 1.
4. The term $8m$ has a coefficient of 8, a base of m, and an exponent of 1.
5. The term $\frac{2}{9}c^7$ has a coefficient of $\frac{2}{9}$, since c^7 is multiplied by $\frac{2}{9}$.

   The base is c and the exponent is 7.

Practice 3

1. $3p$ and $3q$ have different bases, so they are unlike terms.
2. $-k^6$ and $12k^6$ both have a base of k with an exponent of 6, so these terms are like terms.
3. $\frac{3}{8}c^2$ and $8c^2$ both have a base of c with an exponent of 2, so these terms are like terms.
4. $8m^4$ and $8n^4$ have the same exponent, but they have different bases, so they are unlike terms.
5. $5v^{10}$ and $3v^{-10}$ both have a base of v, but the bases have different exponents, so they are unlike terms.

adding and subtracting

I'm an algebra liar. I figure two good lies make a positive.
— TIM ALLEN (1953–   )
AMERICAN ENTERTAINER

In this lesson, you'll learn how to combine a pair of signs into one sign, and how to add and subtract like terms.

WHEN WE ADD two whole numbers or subtract one whole number from another, we only have to work with one sign, either a plus sign or a minus sign. **Integers** are all of the whole numbers, their negatives, and zero. When we add two integers, we often have to work with two or more signs. Sometimes in an integer addition or subtraction problem, two signs appear right next to each other: $2 + -3$.

We can combine a pair of signs into one sign according to the following chart:

Two Signs	Become One Sign
+ +	+
+ −	−
− +	−
− −	+

In other words, if the two signs are different, replace them with the minus sign. If two signs are the same, replace them with a plus sign. $2 + -3$ becomes $2 - 3$, since the two signs between 2 and 3 are different. $4 + (+6)$ becomes $4 + 6$, and $3 - (-7)$ becomes $3 + 7$.

Practice 1

For each problem, combine pairs of signs and solve.

1. $4 + (+10) =$

2. $-3 - (+9) =$

3. $7 - (-1) =$

4. $0 + (-6) =$

5. $9 + (-8) - (-2) =$

6. $15 - (+2) + (-4) =$

7. $-18 + (+20) - (-5) =$

ADDING LIKE TERMS

Remember, like terms have the same base, and those bases have the same exponents. To add two like terms, we add the coefficients of the terms and keep the base and its exponent. Let's say we want to add $4x^3$ and $3x^3$. First, we check that the terms are like terms. Because both have a base of x and an exponent of 3, these are like terms. Next, add the coefficients of the terms. The coefficient of $4x^3$ is 4 and the coefficient of $3x^3$ is 3. $4 + 3 = 7$. Our answer has the same base and the same exponent as the bases and exponents of the terms that we have added, so $4x^3 + 3x^3 = 7x^3$.

TIP: Do not forget to write the base and exponent of your answer. Once you are sure that you are adding like terms, write the base and the exponent of your answer right away. Before finding the sum of $13v^7$ and $28v^7$, write v^7 as part of your answer, and then find the sum of the coefficients.

To find $-8u^8 + (+6u^8)$, we must combine a pair of signs. Because + and + are the same sign, they can be combined into a single plus sign. $-8u^8 + (+6u^8) = -8u^8 + 6u^8$. The base of these terms is u and the exponent is 8, so the base and exponent of our answer is u^8. Add the coefficients: $-8 + 6 = -2$, so $-8u^8 + 6u^8 = -2u^8$.

Practice 2

1. $6a^5 + 2a^5 =$

2. $-2p + 2p =$

3. $23q^{12} + (+11q^{12}) =$

4. $3b^2 + b^2 + 10b^2 =$

5. $10a^3 - (-5a^3) =$

6. $-9t^8 + 8t^8 + (+13t^8) =$

7. $15y^4 + 12y^4 - (-17y^4) =$

SUBTRACTING LIKE TERMS

Just as with addition, to subtract one like term from another, we work with the coefficients of the terms. Our answer has the same base and the same exponent as the term we are subtracting and the term from which we are subtracting. To find the difference between two like terms, subtract the coefficient of the second term from the coefficient of the first term: $8x^7 - 3x^7 = 5x^7$.

What is $-6e^3 - (+4e^3)$? First, check that we have like terms. Both terms have a base of e and an exponent of 3, so we are ready to subtract. Next, combine the two different signs into a single minus sign: $-6e^3 - (+4e^3) = -6e^3 - 4e^3$. Now, subtract 4 from -6: $-6 - 4 = -10$, so $-6e^3 - 4e^3 = -10e^3$.

Practice 3

1. $11g^9 - 9g^9 =$

2. $4j^6 - 5j^6 =$

3. $18n^4 - (+13n^4) =$

4. $h^2 + (-7h^2) =$

5. $-8z^3 - z^3 - z^3 =$

6. $16t^{15} - 9t^{15} - (+t^{15}) =$

7. $3r^{-2} - 4r^{-2} + (-7r^{-2}) =$

ADDING AND SUBTRACTING UNLIKE TERMS

This might be the shortest topic in this book—because you cannot add or subtract unlike terms. The terms $6x$ and $2x^2$ cannot be added, and $2ab$ cannot be subtracted from $10x^3y$. If you see $8y^3 + 2y^6$ or $9m^6 - 4a$, you must leave those terms just as you found them. They cannot be combined. This is why you must always check that you have like terms before adding or subtracting.

Practice 4

For each problem, decide if the terms can be combined or not.

1. $10k^2 - 10k$

2. $n + 19n$

3. $8x^{-3} + (-8x^3)$

4. $-11y^2 + (-15z^2)$

5. $a^5b^4c^2 - (-2a^5b^4c^2)$

ANSWERS

Practice 1

1. Combine the two plus signs into one plus sign and add: $4 + (+10) = 4 + 10 = 14$.
2. Combine the minus sign and the plus sign into a minus sign and subtract: $-3 - (+9) = -3 - 9 = -12$.
3. Combine the two minus signs into one plus sign and add: $7 - (-1) = 7 + 1 = 8$.
4. Combine the plus sign and the minus sign into a minus sign and subtract: $0 + (-6) = 0 - 6 = -6$.
5. Combine the first pair of signs into a minus sign, since the two signs are different: $9 + (-8) - (-2) = 9 - 8 - (-2)$. Combine the second pair of signs into a plus sign, because the two signs are the same: $9 - 8 - (-2) = 9 - 8 + 2 = 1 + 2 = 3$.
6. Combine the first pair of signs into a minus sign, since the two signs are different: $15 - (+2) + (-4) = 15 - 2 + (-4)$. Combine the second pair of signs into a minus sign, because the two signs are also different: $15 - 2 + (-4) = 15 - 2 - 4 = 13 - 4 = 9$.
7. Combine the first pair of signs into a plus sign, since the two signs are the same: $-18 + (+20) - (-5) = -18 + 20 - (-5)$. Combine the second pair of signs into a plus sign, because the two signs are also the same: $-18 + 20 + 5 = 2 + 5 = 7$.

Practice 2

1. Each term has a base of a and an exponent of 5, so the base and exponent of your answer is a^5. Add the coefficients: $6 + 2 = 8$, so $6a^5 + 2a^5 = 8a^5$.
2. Each term has a base of p and an exponent of 1. Remember, if a base does not appear to have an exponent, then it has an exponent of 1. The base of your answer is p. Add the coefficients: $-2 + 2 = 0$, so $-2p + 2p = 0p$, or simply 0.
3. Each term has a base of q and an exponent of 12, so the base and exponent of your answer is q^{12}. Combine the two plus signs into one plus sign: $23q^{12} + (+11q^{12}) = 23q^{12} + 11q^{12}$. Add the coefficients of each term: $23 + 11 = 34$, so $23q^{12} + 11q^{12} = 34q^{12}$.

4. Each term has a base of b and an exponent of 2, so the base and exponent of your answer is b^2. Add the coefficients of each term. Remember, if there is no number before the base of a term, the coefficient of that term is 1: $3 + 1 + 10 = 14$, so $3b^2 + b^2 + 10b^2 = 14b^2$.

5. Each term has a base of a and an exponent of 3, so the base and exponent of your answer is a^3. Combine the two minus signs into one plus sign: $10a^3 - (-5a^3) = 10a^3 + 5a^3$. Add the coefficients of each term: $10 + 5 = 15$, so $10a^3 + 5a^3 = 15a^3$.

6. Each term has a base of t and an exponent of 8, so the base and exponent of your answer is t^8. Combine the two plus signs into one plus sign: $-9t^8 + 8t^8 + (+13t^8) = -9t^8 + 8t^8 + 13t^8$. Add the coefficients of each term: $-9 + 8 + 13 = 12$, so $9t^8 + 8t^8 + 13t^8 = 12t^8$.

7. Each term has a base of y and an exponent of 4, so the base and exponent of your answer is y^4. Combine the two minus signs into one plus sign: $15y^4 + 12y^4 - (-17y^4) = 15y^4 + 12y^4 + 17y^4$. Add the coefficients of each term: $15 + 12 + 17 = 44$, so $15y^4 + 12y^4 + 17y^4 = 44y^4$.

Practice 3

1. Each term has a base of g and an exponent of 9, so the base and exponent of your answer is g^9. Subtract the coefficient of the second term from the coefficient of the first term: $11 - 9 = 2$, so $11g^9 - 9g^9 = 2g^9$.

2. Each term has a base of j and an exponent of 6, so the base and exponent of your answer is j^6. Subtract the coefficient of the second term from the coefficient of the first term: $4 - 5 = -1$, so $4j^6 - 5j^6 = -1j^6$, or simply $-j^6$.

3. Each term has a base of n and an exponent of 4, so the base and exponent of your answer is n^4. Combine the two different signs into one minus sign: $18n^4 - (+13n^4) = 18n^4 - 13n^4$. Subtract the coefficient of the second term from the coefficient of the first term: $18 - 13 = 5$, so $18n^4 - 13n^4 = 5n^4$.

4. Each term has a base of h and an exponent of 2, so the base and exponent of your answer is h^2. Combine the two different signs into one minus sign: $h^2 + (-7h^2) = h^2 - 7h^2$. Subtract the coefficient of the second term from the coefficient of the first term. There is no number before the base of the h^2 term, which means that the coefficient of that term is 1: $1 - 7 = -6$, so $h^2 - 7h^2 = -6h^2$.

5. Each term has a base of z and an exponent of 3, so the base and exponent of your answer is z^3. There is no number before the base of two of the z^3 terms, which means that the coefficient of both terms is 1. Next, work with the coefficients. Subtract 1 twice from –8: $-8 - 1 - 1 = -10$, so $-8z^3 - z^3 - z^3 = -10z^3$.

6. Each term has a base of t and an exponent of 15, so the base and exponent of your answer is t^{15}. Combine the two different signs into one minus sign: $16t^{15} - 9t^{15} - (+t^{15}) = 16t^{15} - 9t^{15} - t^{15}$. There is no number before the base of the last term, which means that the coefficient of that term is 1. Next, work with the coefficients. Subtract 9 and 1 from 16: $16 - 9 - 1 = 6$, so $16t^{15} - 9t^{15} - t^{15} = 6t^{15}$.

7. Each term has a base of r and an exponent of –2, so the base and exponent of your answer is r^{-2}. Combine the two different signs into one minus sign: $3r^{-2} - 4r^{-2} + (-7r^{-2}) = 3r^{-2} - 4r^{-2} - 7r^{-2}$. Now, work with the coefficients. Subtract 4 and 7 from 3: $3 - 4 - 7 = -8$, so $3r^{-2} - 4r^{-2} - 7r^{-2} = -8r^{-2}$.

Practice 4

1. $10k^2$ and $10k$ have the same base, but different exponents, so $10k$ cannot be subtracted from $10k^2$.

2. n and $19n$ have the same base (n) and the same exponent (1), so n can be added to $19n$.

3. $8x^{-3}$ and $-8x^3$ have the same base (x), but different exponents; 3 and –3 are not the same exponent, so $8x^{-3}$ and $-8x^3$ cannot be combined.

4. $-11y^2$ and $-15z^2$ have the same exponent (2) but different bases, so $-11y^2$ and $-15z^2$ cannot be combined.

5. $a^5b^4c^2$ and $-2a^5b^4c^2$ have the same bases (a, b, and c) and the same exponents for each base (5 for a, 4 for b, and 2 for c), so $a^5b^4c^2$ and $-2a^5b^4c^2$ can be combined.

multiplying and dividing

Can you do Division? Divide a loaf by a knife—
what's the answer to that?
—LEWIS CARROLL (1832–1898)
ENGLISH AUTHOR AND MATHEMATICIAN

In this lesson, you'll learn how to multiply and divide like and unlike terms.

NOW THAT YOU KNOW how to add and subtract terms, let's look at how to multiply and divide them. While unlike terms cannot be combined (added or subtracted), we can multiply and divide both like terms and unlike terms. In fact, multiplication and division of unlike terms is no different from multiplication and division of like terms!

Just as with addition and subtraction though, we must first review signed numbers. What happens when a positive factor is multiplied by a negative factor? Or when two negative factors are multiplied? The following chart might look familiar.

Sign of One Factor	Sign of Second Factor	Sign of Product
+	+	+
+	−	−
−	+	−
−	−	+

When two factors of the same sign are multiplied, the **product** is positive. A product is the result when you are multiplying two factors. When two factors of different signs are multiplied, the result is negative.

....

TIP: This chart is similar to the chart for combining plus and minus signs in Lesson 2. Always remember: Two of the same sign, think positive; two different signs, think negative.

....

The division chart looks just like the multiplication chart.

Sign of the Dividend	Sign of the Divisor	Sign of the Quotient
+	+	+
+	−	−
−	+	−
−	−	+

In a division problem, the term being divided is called the *dividend*, the term doing the dividing is called the *divisor*, and the result of the division is called the *quotient*.

Practice 1

1. $(5)(6) =$

2. $(-4)(9) =$

3. $(24) \div (-6) =$

4. $(-32) \div (-8) =$

MULTIPLYING TERMS

When we added two like terms, we worked only with the coefficients. The base and exponent of our sum was the same as the base and exponent of each **addend** (an addend is a term that is added to another term). To multiply two terms, we must work with the coefficients, the bases, and the exponents.

First, multiply the coefficients. This product is the coefficient of our answer. Next, put every base in either term into the base of our answer. Finally, add the exponents of the bases that each term has in common, and those sums are the exponents of each base in the answer. Confused? Let's look at an example.

Example
What is the product of $2m^3$ and $5m^5$?

First, multiply the coefficients: $(2)(5) = 10$.
Next, put every base in either term into the answer. Both terms have a base of m, so the base of our answer is m.
Finally, add the exponents of the bases that each term has in common. In other words, add the exponent of m in $2m^3$ to the exponent of m in $5m^5$. The exponent of m in $2m^3$ is 3 and the exponent of m in $5m^5$ is 5. $3 + 5 = 8$. The exponent of m in our answer is 8.
Now, put it all together. The coefficient of our answer is 10, the base of our answer is m, and the exponent of our answer is 8: $10m^8$. The product of $2m^3$ and $5m^5$ is $10m^8$.

Multiplying can be a bit trickier when the bases of each factor are a little different. Here's another example.

Example
$(8x^2y^3)(-9x^7) =$

Follow the same steps. First, multiply the coefficients: $(8)(-9) = -72$.
Put every base in either term into the answer. Both terms have a base of x, so x is a base in our answer. The first term also has a base of y, so our answer has a base of y, too.
Next, add the exponents of the bases that each term has in common. The exponent of x in $8x^2y^3$ is 2 and the exponent of x in $-9x^7$ is 7, so the exponent of x in our answer is $2 + 7 = 9$. The exponent of y in $8x^2y^3$ is 3. y does not appear at all in the second term, so the exponent of y in our answer is 3.
Finally, put it all together: the coefficient of our answer is -72, one base of our answer is x with an exponent of 9, and the other base of our answer is y with an exponent of 3: $-72x^9y^3$.

Multiplying terms with completely different bases is actually easiest of all. Because the terms have no bases in common, there are no exponents to add. Just copy the bases and exponents of each term right into your answer.

Example

$(-12a^3bc^6)(-9xy^9z^5) =$

Multiply the coefficients: $(-12)(-9) = 108$.

Put every base in either term into the answer. Our answer has bases of a, b, c, x, y, and z.

Because none of the bases in the first term appear in the second term, we have no exponents to add. The exponents of the bases in our answer are the same as the exponents of the bases in each term. Our answer is $108a^3bc^6xy^9z^5$.

Practice 2

1. $(-7x^{10})(4x^6) =$

2. $(14u)(8u^4) =$

3. $(-3y^9)(-y^9) =$

4. $(10g^3h^5)(-2g^5h^9) =$

5. $(9a^6b^{11}c^4)(9a^2c^2) =$

6. $(-5j^7k^7l^4)(11l^4mn^9) =$

7. $(-10p^6)(-12qr^{10})$

DIVIDING TERMS

We follow some similar steps to divide terms. When multiplying two terms, we began by multiplying coefficients. To divide two terms, first, divide the coefficient of the first term by the coefficient of the second term. For bases that are in the dividend but not the divisor, carry those bases and their exponents into the answer. The next step might surprise you: If there are any bases in the divisor that are NOT in the dividend, carry them into our answer, too, but change the sign of their exponents. Finally, for each base that is common to both the dividend and the divisor, subtract the exponent of the base in the divisor from the exponent of the base in our dividend. That is the exponent of that base in our answer. Whew! Is it really that difficult? Not after you see a few examples.

Example

$(8a^5) \div (2a^3) =$

First, divide the coefficient of the dividend by the coefficient of the divisor: $8 \div 2 = 4$.

Next, carry the bases of the dividend into your answer. Your answer has a base of a.

There are no bases in the dividend that are not in the divisor, and vice versa, so move right on to the next step.

For each base that is common to both the dividend and the divisor, subtract the exponent of the base in the divisor from the exponent of the base in the dividend. The exponent of a in $8a^5$ is 5 and the exponent of a in $2a^3$ is 3; $5 - 3 = 2$. The exponent of a in our answer is 2. $(8a^5) \div (2a^3) = 4a^2$.

Now, let's see an example where the dividend has a base that the divisor does not have, and the divisor has a base that the dividend does not have.

Example

$(35g^{10}) \div (5y^4) =$

Divide the coefficient of the dividend by the coefficient of the divisor: $35 \div 5 = 7$.

Carry the bases of the dividend into the answer with its exponent, since that base is not present in the divisor. The answer has a base of g with an exponent of 10.

Carry the bases that are in the divisor but not the dividend into the answer. Change the sign of their exponents. The divisor has a base of y that is not in the dividend, so y will be in the answer, but with an exponent of -4 instead of 4.

There are no terms common to the dividend and the divisor, so we have no subtraction to do. $(35g^{10}) \div (5y^4) = 7g^{10}y^{-4}$.

Why did the exponent of y change from positive to negative? Because we wanted to divide the dividend by y, but there was no y in the dividend. That means we were unable to divide by y, and we need to show in our answer that this division was never performed. A negative exponent means that if we were to write a term as a fraction, the bases with negative exponents would be written in the denominator of the fraction (and their exponents would change to positive). Remember, fractions mean division, so if a term appears in the denominator of a fraction, that means the term is acting like a divisor. Our answer, $7g^{10}y^{-4}$, could also be written as $\frac{7g^{10}}{y^4}$. In fact, we could write both the original problem and the answer as fractions: $\frac{35g^{10}}{5y^4} = \frac{7g^{10}}{y^4}$.

Now, it's a little easier to see what happened: 35 was divided by 5, g^{10} was not divided by anything and went straight into our answer, and we could not divide by y^4, so we kept it in the denominator of our answer.

You probably still have a few questions. What if a base is common to the dividend and the divisor, but the exponent of the divisor is greater? Or, what if a base is common to the dividend and the divisor, and their exponents are the same? Let's look at both of these cases.

Example
$(18a^4b^3) \div (-6ab^5) =$

Begin with the coefficients: $18 \div -6 = -3$.

Carry the bases of the dividend into the answer. The answer has bases of a and b.

There are no bases in the dividend that are not in the divisor, and vice versa, so we are ready to subtract.

The exponent of a in the dividend is 4 and the exponent of a in the divisor is 1: $4 - 1 = 3$. The exponent of a in the answer is 3. The exponent of b in the dividend is 3 and the exponent of b in the divisor is 5. $3 - 5 = -2$. The exponent of b in the answer is -2.

Put it all together: $(18a^4b^3) \div (-6ab^5) = -3a^3b^{-2}$. It is okay to have a negative exponent in your answer! If you would like to avoid negative exponents, this answer can also be written as $\frac{-3a^3}{b^2}$.

Example
$(16c^7) \div (4c^7) =$

Begin again with the coefficients: $16 \div 4 = 4$.

There is only one base in the dividend and the divisor, c. Subtract the exponent of c in the dividend from the exponent of c in the denominator: $7 - 7 = 0$. What does this mean? We could write our answer as $4c^0$, but because any number or variable to the power of 0 is equal to 1, $c^0 = 1$ and $4(1) = 4$. In other words, if the exponent of a base is the same in both the dividend and the divisor, that base does not appear in your answer.

....

TIP: If the coefficient of the dividend and the divisor are the same, your answer does not appear to have a coefficient. That is because any value divided by itself is equal to 1, and it is unnecessary to write the number 1 as the coefficient of a term.

....

Practice 3

1. $(15v^9) \div (3v^5) =$

2. $(-63a) \div (-7a^8) =$

3. $(30p^{11}q^4r^2) \div (-5q^2r^3) =$

4. $(11x^3y^5) \div (11y^5z^2) =$

5. $(-42m^2n^{-2}o^5) \div (6mn^2) =$

ANSWERS

Practice 1

1. Both factors are positive, so your answer is positive: $(5)(6) = 30$.
2. One factor is positive and one is negative, so your answer is negative: $(-4)(9) = -36$.
3. One number is positive and one is negative, so your answer is negative: $(24) \div (-6) = -4$.
4. Both numbers are negative, so your answer is positive: $(-32) \div (-8) = 4$.

Practice 2

1. Multiply the coefficients: $(-7)(4) = -28$. Each term has a base of x, so your answer has a base of x. Next, add the exponents of x from each term: $10 + 6 = 16$. $(-7x^{10})(4x^6) = -28x^{16}$.

2. Multiply the coefficients: $(14)(8) = 112$. Each term has a base of u, so your answer has a base of u. Next, add the exponents of u from each term. Remember, if a base appears to have no exponent, then it has an exponent of 1. $1 + 4 = 5$. $(14u)(8u^4) = 112u^5$.

3. Multiply the coefficients: $(-3)(-1) = 3$. Each term has a base of y, so your answer has a base of y. Next, add the exponents of y from each term: $9 + 9 = 18$. $(-3y^9)(-y^9) = 3y^{18}$.

4. Multiply the coefficients: $(10)(-2) = -20$. Each term has a base of g and a base of h, so your answer has both g and h in it. Next, add the exponents of g from each term: $3 + 5 = 8$. The exponent of g in your answer is 8. Finally, add the exponents of h from each term: $5 + 9 = 14$. The exponent of h in your answer is 14. $(10g^3h^5)(-2g^5h^9) = -20g^8h^{14}$.

5. Multiply the coefficients: $(9)(9) = 81$. The first term has bases of a, b, and c (the second term also has bases of a and c), so your answer has bases of a, b, and c. Next, add the exponents of a from each term: $6 + 2 = 8$. The exponent of a in your answer is 8. Because the second term does not contain b, the exponent of b in your answer will be the same as the exponent of b in the first term, 11. Finally, add the exponents of c from each term: $4 + 2 = 6$. The exponent of c in your answer is 6. $(9a^6b^{11}c^4)(9a^2c^2) = 81a^8b^{11}c^6$.

6. Multiply the coefficients: $(-5)(11) = -55$. The first term has bases of j, k, and l and the second term has bases of l, m, and n, so your answer has bases of j, k, l, m, and n. Next, add the exponents of l from each term: $4 + 4 = 8$. The exponent of l in your answer is 8. Finally, because the bases j, k, l, m, and n each appear in only one term, their exponents are carried right into your answer: $(-5j^7k^7l^4)(11l^4mn^9) = -55j^7k^7l^8mn^9$.

7. Multiply the coefficients: $(-10)(-12) = 120$. The first term has a base of p and the second term has bases of q and r, so your answer has bases of p, q, and r. Because the terms have no bases in common, there are no exponents to add. The exponents of p, q, and r are carried right into your answer. $(-10p^6)(-12qr^{10}) = 120p^6qr^{10}$.

Practice 3

1. Begin with the coefficients: $15 \div 3 = 5$. Carry the base (v) from the dividend into the answer. There are no bases in the divisor that are not in the dividend, and vice versa. Next, subtract the exponent of v in the divisor from the exponent of v in the dividend: $9 - 5 = 4$. The exponent of v in the answer is 4. $(15v^9)(3v^5) = 5v^4$.

2. Begin with the coefficients: $-63 \div -7 = 9$. Carry the base (a) from the dividend into the answer. There are no bases in the divisor that are not in the dividend, and vice versa. Next, subtract the exponent of a in the divisor from the exponent of a in the dividend: $1 - 8 = -7$. The exponent of a in the answer is -7. $(-63a)(-7a^8) = 9a^{-7}$.

3. Begin with the coefficients: $30 \div -5 = -6$. Carry the bases (p, q, and r) from the dividend into the answer. There are no bases in the divisor that are not in the dividend. Because p does not appear in the divisor, carry its exponent into the answer. Next, subtract the exponent of q in the divisor from the exponent of q in the dividend: $4 - 2 = 2$. Finally, subtract the exponent of r in the divisor from the exponent of r in the dividend: $2 - 3 = -1$. The exponent of r in the answer is -1. $(30p^{11}q^4r^2)(-5q^2r^3) = -6p^{11}q^2r^{-1}$.

4. Begin with the coefficients: $11 \div 11 = 1$. Carry the bases (x and y) from the dividend into the answer. The divisor has a base (z) that is not in the dividend, so next, carry it into the answer. Change its exponent from 2 to -2. Because x does not appear in the divisor, carry its exponent into the answer. Finally, subtract the exponent of y in the divisor from the exponent of y in the dividend: $5 - 5 = 0$, so our answer does not have a base of y after all. $(11x^3y^5)(11y^5z^2) = x^3z^{-2}$.

5. Begin with the coefficients: $-42 \div 6 = -7$. Carry the bases (m, n, and o) from the dividend into the answer. Because o does not appear in the divisor, carry its exponent into the answer. Next, subtract the exponent of m in the divisor from the exponent of m in the dividend: $2 - 1 = 1$. The exponent of m in the answer is 1. Finally, subtract the exponent of n in the divisor from the exponent of n in the dividend: $-2 - 2 = -4$. The exponent of n in the answer is -4. $(-42m^2n^{-2}o^5)(6mn^2) = -7mn^{-4}o^5$.

LESSON 4

single-variable expressions

"Obvious" is the most dangerous word in mathematics.
—ERIC TEMPLE BELL (1883–1960)
MATHEMATICIAN AND SCIENCE FICTION AUTHOR

In this lesson, you'll review the order of operations and learn how to evaluate algebraic expressions.

THE ADDITION SENTENCE $3x + 7x$, or even just $3x$ alone is an algebraic expression. An **algebraic expression** is one or more terms, at least one of which contains a variable, which may or may not contain an operation (such as addition or multiplication).

We have seen how to add, subtract, multiply, and divide terms, but all of our answers have contained variables. To evaluate an expression, we replace variables with real numbers. At first glance, it might seem easy to evaluate an expression once the variables have been replaced with numbers, but we must remember to follow the order of operations, or we will arrive at the wrong answer.

The order of operations is a list that tells us how to go about evaluating an expression. First, handle any operations that are in parentheses, no matter what those operations are. Next, work with the exponents in the expression. After that, you can do multiplication and division. Finally, perform addition and subtraction. Most people use the acronym PEMDAS to help them remember the order of operations:

P	Parentheses	**P**	Please
E	Exponents	**E**	Excuse
M	Multiplication	**M**	My
D	Division	**D**	Dear
A	Addition	**A**	Aunt
S	Subtraction	**S**	Sally

On the left is a list of the order of operations, and on the right is a phrase (*Please Excuse My Dear Aunt Sally*) that can help you remember the order of operations. The first letter in each word of the phrase is the same as the first letter in each operation.

To evaluate the numerical expression $6(5) + 4$, we multiply 6 and 5 first, since multiplication comes before addition in the order of operations: $6(5) = 30$. Then, we add: $30 + 4 = 34$. If we were to add 5 and 4 first and then multiply by 6, our answer would be 54, which is incorrect. That's why the order of operations is so important.

Example

$2(6 + 4) - 4^2 =$

Begin with the operation in parentheses: $6 + 4 = 10$. The expression is now $2(10) - 4^2$.

Next, work with the exponents: $4^2 = 16$, and the expression becomes $2(10) - 16$.

Multiplication is next: $2(10) = 20$, and we are left with $20 - 16$.

Finally, subtract: $20 - 16 = 4$. The expression $2(6 + 4) - 4^2$ is equal to 4.

TIP: If there is more than one operation inside a set of parentheses, use the order of operations to tell you which operation to perform first. In the expression $(5 + 4(3)) - 2$, addition and multiplication are both inside parentheses. Because multiplication comes before addition in the order of operations, we begin by multiplying 4 and 3.

Practice 1

Evaluate each expression.

1. $5 - 6 \div 2$

2. $-2(10 + 3)$

3. $3^2(7) + 1$

4. $(-5 + 12)(21 - 13)$

5. $(10 - (8 - 2^2))3$

EVALUATING SINGLE-VARIABLE EXPRESSIONS

Now that we know how to handle many operations in an expression, we can look at how to evaluate algebraic expressions. The most basic algebraic expressions contain only one variable, and these are called single-variable expressions. When we are given a value for the variable, we can replace that variable with the value (in parentheses).

Example
What is $6x$ when $x = 4$?

Replace x in the expression $6x$ with 4: $6(4) = 24$. When $x = 4$, $6x = 24$. But what if $x = -10$? Then we would replace x with -10: $6(-10) = -60$. The value of $6x$ varies depending on the value of x. This is why x is called a variable!

Example
What is $-5p + 9$ when $p = -1$?

Replace p with -1: $-5(-1) + 9$. Remember the order of operations: Multiply before adding. $-5(-1) = 5$, $5 + 9 = 14$. When $p = -1$, $-5p + 9$ is equal to 14.

Example
What is $4(c^2 - 7)$ when $c = -3$?

Replace c with -3: $4((-3)^2 - 7)$. $(-3)^2 - 7$ is in parentheses, and since exponents come before subtraction, begin there. $(-3)^2 = 9$, $9 - 7 = 2$. Finally, $4(2) = 8$.

Practice 2

1. What is $-9z$ when $z = 9$?

2. What is $4u - 3$ when $u = 7$?

3. What is $-8 + 2g$ when $g = -5$?

4. What is $\frac{1}{4}k + 30$ when $k = 20$?

5. What is $12(4 - x)$ when $x = 1$?

6. What is $q^2 + 15$ when $q = -6$?

7. What is $-8m^2$ when $m = 4$?

8. What is $4d \div 3 - 20$ when $d = 12$?

9. What is $(2b + 1)^2$ when $b = -2$?

10. What is $-5(3a^2 - 24)$ when $a = 3$?

SIMPLIFYING EXPRESSIONS

Some expressions may have only one variable, but that variable appears more than once in the expression. For instance, $8x + 2x$, or $2(j - 6) - 7j$. We have two choices for evaluating these expressions. We can replace every occurrence, or instance, of the variable with its value, or we can simplify the expression first and then replace the variable with its value.

Example
What is $5v - 8v$ when $v = 10$?

First, try replacing every v with 10: $5(10) - 8(10)$.

Multiplication comes before subtraction, so multiply first: $5(10) = 50$ and $8(10) = 80$.

The expression is now $50 - 80$.

Subtract: $50 - 80 = -30$.

Now, evaluate the expression again, only this time, simplify it first. Remember, to subtract like terms, subtract the coefficient of the second term from the coefficient of the first term, and keep the exponent and base: $5v - 8v = -3v$.

Replace v with 10 and multiply: $-3(10) = -30$.

Either way, we arrive at the same answer. Which way was easier? That is up to you. You might find simplifying first easier, or you might prefer to just substitute the variables with their values right away. As long as you follow the order of operations, both methods will work.

....

TIP: Some expressions cannot be simplified, or can be only partially simplified, often because they do not contain like terms. To evaluate $3x + 3x^2$ when $x = 1$, replace both x's with 1 because $3x$ and $3x^2$ cannot be combined.

....

Practice 3

1. What is $y + 11y$ when $y = 6$?

2. What is $12n - 8n + 7$ when $n = 9$?

3. What is $2r^2 + 3r^2$ when $r = -3$?

4. What is $4h - 8h + 6(h + 2)$ when $h = 4$?

5. What is $5s + s^2 - 9$ when $s = -2$?

ANSWERS

Practice 1

1. This expression contains subtraction and division. Since division comes before subtraction in the order of operations, divide first: $6 \div 2 = 3$.
   The expression becomes $5 - 3$.
   Subtract: $5 - 3 = 2$.
2. Parentheses are first in the order of operations.
   $10 + 3 = 13$, and the expression becomes $-2(13)$.
   Multiply: $-2(13) = -26$.
3. This expression contains an exponent, multiplication, and addition. Exponents come before multiplication and addition, so begin with 3^2: $3^2 = 9$.
   The expression is now $9(7) + 1$.
   Multiplication comes before addition, so multiply next:
   $9(7) = 63$, and the expression becomes $63 + 1$.
   Finally, add: $63 + 1 = 64$.
4. There are two sets of parentheses in this expression, so work on each of them separately.
   The first set of parentheses contains addition: $-5 + 12 = 7$.
   The second set of parentheses contains subtraction: $21 - 13 = 8$.
   The expression is now $(7)(8)$.
   Multiply: $(7)(8) = 56$.
5. The left side of the expression contains parentheses within parentheses, so start with the innermost parentheses: $(8 - 2^2)$.
   Because exponents come before subtraction, start with the exponent:
   $2^2 = 4$, and the parentheses become $(8 - 4)$.
   Subtract: $8 - 4 = 4$.
   The expression is now $(10 - 4)3$.
   The subtraction is in parentheses, so handle it before multiplying:
   $10 - 4 = 6$. The expression becomes $(6)3$.
   Finally, multiply: $(6)3 = 18$.

Practice 2

1. Replace z with 9 and multiply:
   $$-9(9) = -81$$
2. Replace u with 7:
   $$4(7) - 3$$

Multiplication comes before subtraction in the order of operations, so multiply next:

$$4(7) = 28$$

The expression becomes $28 - 3$.

Subtract: $28 - 3 = 25$.

3. Replace g with -5:

$$-8 + 2(-5)$$

Multiply before adding:

$$2(-5) = -10$$

The expression becomes $-8 + -10$.

Add: $-8 + -10 = -18$.

4. Replace k with 20:

$$\tfrac{1}{4}(20) + 30$$

Multiply before adding:

$$\tfrac{1}{4}(20) = 5$$

The expression becomes $5 + 30$.

Add: $5 + 30 = 35$.

5. Replace x with 1:

$$12(4 - 1)$$

Subtraction is in parentheses, so subtract before multiplying:

$$(4 - 1) = 3$$

The expression becomes $12(3)$.

Multiply: $12(3) = 36$.

6. Replace q with -6:

$$(-6)^2 + 15$$

Exponents come before addition in the order of operations, so handle the exponent first:

$$(-6)^2 = 36$$

The expression becomes $36 + 15$.

Add: $36 + 15 = 51$.

7. Replace m with 4:

$$-8(4)^2$$

Exponents come before multiplication, so handle the exponent first:

$$4^2 = 16$$

The expression becomes $-8(16)$.

Multiply: $-8(16) = -128$.

8. Replace d with 12:

$$4(12) \div 3 - 20$$

Multiply first:

$$4(12) = 48$$

The expression becomes $48 \div 3 - 20$.

Division comes before subtraction:

$48 \div 3 = 16$

We are left with $16 - 20$.

Subtract: $16 - 20 = -4$.

9. Replace b with -2:

$(2(-2) + 1)^2$

There is multiplication and addition inside the parentheses. Multiply first:

$2(-2) = -4$

The expression becomes $(-4 + 1)^2$.

Because the addition is in parentheses, it must be done before the exponent is handled:

$-4 + 1 = -3$

We are left with $(-3)^2$.

Finally, square -3: $(-3)^2 = 9$.

10. Replace a with 3:

$-5(3(3)^2 - 24)$

There is multiplication, an exponent, and subtraction inside the parentheses. Because exponents come before multiplication and subtraction, handle the exponent first:

$(3)^2 = 9$

The expression becomes $-5(3(9) - 24)$.

Multiplication inside the parentheses comes next:

$3(9) = 27$

The expression becomes $-5(27 - 24)$.

The subtraction is in parentheses, so it must be done before the multiplication:

$27 - 24 = 3$

We are left with $-5(3)$.

Multiply: $-5(3) = -15$.

Practice 3

1. Add y and $11y$ by adding the coefficients and keeping the base and exponent: $1 + 11 = 12$, so $y + 11y = 12y$.

Replace y with 6 and multiply:

$12(6) = 72$

2. Subtract $8n$ from $12n$. Remember, subtract the coefficient of $8n$ from the coefficient of $12n$ and keep the base and exponent:

$$12n - 8n = 4n$$

The expression is now $4n + 7$.

Replace n with 9:

$$4(9) + 7$$

Multiplication comes before addition in the order of operations:

$$4(9) = 36$$

The expression is now $36 + 7$.

Add: $36 + 7 = 43$.

3. Add $2r^2$ and $3r^2$ by adding the coefficients and keeping the base and exponent:

$$2r^2 + 3r^2 = 5r^2$$

Replace r with -3:

$$5(-3)^2$$

Exponents come before multiplication in the order of operations:

$$(-3)^2 = 9$$

The expression is now $5(9)$.

Multiply: $5(9) = 45$.

4. Replace every h with 4:

$$4(4) - 8(4) + 6((4) + 2)$$

Begin with addition inside the parentheses:

$$4 + 2 = 6$$

The expression is now:

$$4(4) - 8(4) + 6(6)$$

Multiplication comes before subtraction or addition, so multiply next:

$$4(4) = 16$$
$$8(4) = 32$$
$$6(6) = 36$$

The expression is now:

$$16 - 32 + 36$$

Add -32 and 36:

$$-32 + 36 = 4$$

The expression is now:

$$16 + 4$$

Add: $16 + 4 = 20$.

5. None of the terms in the expression $5s + s^2 - 9$ can be combined because they are unlike terms. Replace every s with -2:

$$5(-2) + (-2)^2 - 9$$

Work with the exponent first:

$(-2)^2 = 4$

The expression is now:

$5(-2) + 4 - 9$

Multiplication comes before addition or subtraction, so multiply next:

$5(-2) = -10$

The expression is now:

$-10 + 4 - 9$

You can add before subtracting (or you can subtract before adding):

$-10 + 4 = -6$

The expression is now:

$-6 - 9$

Finally, subtract: $-6 - 9 = -15$.

LESSON

multivariable expressions

The methods which I set forth do not require either constructions or geometrical or mechanical reasonings: but only algebraic operations, subject to a regular and uniform rule of procedure.
—JOSEPH-LOUIS LAGRANGE (1736–1813)
ITALIAN MATHEMATICIAN AND ASTRONOMER

In this lesson, you'll learn how to simplify and evaluate multivariable expressions.

THE PREVIOUS LESSON showed us that we do not need to simplify single-variable expressions. We can just substitute the value for the variable and evaluate. The same is true for multivariable expressions (expressions with more than one different variable), but you might find yourself performing many calculations if you do not simplify first.

To simplify a multivariable expression, we add, subtract, multiply, or divide all of the like terms until we are left with only unlike terms. Sometimes, we might take three, four, five, or even more steps to simplify an expression. Other times, there might not be a single operation we can perform. This occurs when all of the terms are unlike right from the start.

Example
Simplify $3x^2 + 5y + 2x^2 - 3y$.

This expression has two x^2 terms and two y terms. We can simplify the expression into just two terms by combining the two x^2 terms and the two y

terms: $3x^2 + 2x^2 = 5x^2$ and $5y - 3y = 2y$. Therefore, $3x^2 + 5y + 2x^2 - 3y$ simplifies to $5x^2 + 2y$.

Example
Simplify $2a^3 - 4a + 5 + 9a - 6b$.

There is only one a^3 term, only one b term, and only one constant (5). None of these terms can be combined with any other terms. The only terms that can be combined are the a terms, $-4a$ and $9a$; $-4a + 9a = 5a$, so $2a^3 - 4a + 5 + 9a - 6b$ simplifies to $2a^3 + 5 + 5a - 6b$. Remember, $2a^3$ and $5a$ are unlike terms. Although they have the same base, those bases have different exponents.

Example
Simplify $4w - 6t + 2s$.

These terms are all unlike, because they all have different bases. This expression is already in its simplest form.

....

TIP: When you think you have finished simplifying an expression, write all the terms that contain the same variable next to each other to be sure that there are no terms left that can be combined.

....

Example
Simplify $-7gh + 8g - 2h + 5gh$.

This expression contains three kinds of terms: terms with a base of g, terms with a base of h, and terms with a base of gh. The two terms with a base of gh can be combined: $-7gh + 5gh = -2gh$. Then, $-7gh + 8g - 2h + 5gh$ simplifies to $-2gh + 8g - 2h$.

Example
Simplify $6x^2 + 2x(x - 4y) + 3xy$.

At first glance, this expression appears to contain four kinds of terms: x^2 terms, x terms, y terms, and xy terms. However, the first step in simplifying this expression is to multiply: $2x(x - 4y)$ can be simplified using the distributive law. The **distributive law** says that $a(b + c) = ab + ac$. In words, the law says that the term outside the parentheses should be multiplied by each term inside the parentheses. To find $2x(x - 4y)$, multiply $2x$ by x and by $-4y$: $2x(x) = 2x^2$ and $2x(-4y) = -8xy$. Our expression is now $6x^2 + 2x^2 - 8xy + 3xy$. This expression has two x^2 terms that can be combined and two xy terms that can be combined: $6x^2 + 2x^2 = 8x^2$ and $-8xy + 3xy = -5xy$. The expression $6x^2 + 2x(x - 4y) + 3xy$ simplifies to $8x^2 - 5xy$.

Practice 1

Simplify each expression.

1. $x + 5y - 3x$

2. $2p - 9p^2 + 8r^2 + 2p^2$

3. $7s - (4s + 2t) + 10t$

4. $2b^3 + 6c - b^3 - 5c$

5. $j + k + l - 2j - 3k - 4l$

6. $2(x + y) - 5x$

7. $wv + w(v - 1)$

8. $6f^5g + 3f^2 - 4g - 3f^5g + g$

9. $-9a^2 + 5(2b + 3) - b$

10. $4(k + m) - 9(3k - 2m) + k + m$

EVALUATING MULTIVARIABLE EXPRESSIONS

Now that we know how to simplify multivariable expressions, we can evaluate them. To evaluate a multivariable expression, replace each variable in the expression with its value. As always, remember the order of operations!

> **Example**
> What is $5y - 2z$ when $y = 3$ and $z = -1$?

The terms $5y$ and $-2z$ are unlike, so they cannot be simplified. Replace y with 3 and replace z with -1:

$$5(3) - 2(-1) = 15 + 2 = 17$$

Example

What is $3a + 4b - 4a + 3b$ when $a = 5$ and $b = 6$?

First, combine the two a terms and the two b terms:

$$3a - 4a = -a$$
$$4b + 3b = 7b$$

The expression is now $-a + 7b$. Replace a with 5 and b with 6:

$$-5 + 7(6) = -5 + 42 = 37$$

Example

What is $3(s + 2t) - 6t + 4s$ when $s = -2$ and $t = 32$?

First, use the distributive law to find $3(s + 2t)$. Multiply 3 by s and multiply 3 by $2t$:

$$3(s + 2t) = 3s + 6t$$

The expression is now:

$$3s + 6t - 6t + 4s$$

Combine the two s terms and the two t terms:

$$3s + 4s = 7s$$
$$6t - 6t = 0$$

The expression is now just $7s$. Replace s with -2. There are no t terms to replace.

$$7(-2) = -14$$

The preceding example shows why simplifying an expression can often make evaluating an expression easier. Rather than having to add, subtract, multiply, or divide with 32, the value of t, in the end, the only calculation we had to do after substituting the value of s was to multiply 7 by -2.

Practice 2

Find the value of each expression when $r = 3$ and $t = -2$.

1. $3r^2 + 4t$

2. $-r + 5t - 2t + 9r$

3. $6r - 3t^2 + 4t - 2r$

4. $4rt - 2r(8 + t)$

Find the value of each expression when $a = -3$, $b = 6$, and $c = 1$.

5. $2a + 3b - 4c - (5a + 4b)$

6. $4(a - b + 2c) + c(a - 3)$

7. $3c^2 - 3ab + b(3a + c)$

8. $5b + 2b(c - a) + 3ab$

Find the value of each expression when $w = 1$, $x = -2$, $y = 3$, and $z = -4$.

9. $6(3w + x) + 2y - 3(w - z)$

10. $w^2 - 3x + 5(z + 4) - 10(\frac{1}{2}z + y)$

ANSWERS

Practice 1

1. This expression has two x terms and one y term.
   Combine the x terms: $x - 3x = -2x$.
   The y term cannot be combined with anything, so $x + 5y - 3x$ simplifies to $-2x + 5y$.
2. This expression has one p term, two p^2 terms, and one r^2 term.
   Combine the p^2 terms: $-9p^2 + 2p^2 = -7p^2$.
   The other terms cannot be combined, so $2p - 9p^2 + 8r^2 + 2p^2$ simplifies to $2p - 7p^2 + 8r^2$.
3. This expression has two s terms and two t terms.
   Combine the s terms: $7s - 4s = 3s$.
   Combine the t terms. Remember: the $2t$ term is subtracted, so it is negative: $-2t + 10t = 8t$.
   $7s - (4s + 2t) + 10t$ simplifies to $3s + 8t$.
4. This expression has two b^3 terms and two c terms.
   Combine the b^3 terms: $2b^3 - b^3 = b^3$.

Combine the c terms: $6c - 5c = c$.

$2b^3 + 6c - b^3 - 5c$ simplifies to $b^3 + c$.

5. This expression has two j terms, two k terms, and two l terms.

Combine the j terms: $j - 2j = -j$.

Combine the k terms: $k - 3k = -2k$.

Combine the l terms: $l - 4l = -3l$.

$j + k + l - 2j - 3k - 4l$ simplifies to $-j - 2k - 3l$.

6. This expression has two x terms and one y term before we multiply.

Use the distributive law to find $2(x + y)$. Multiply 2 by x and multiply 2 by y:

$$2(x + y) = 2x + 2y$$

The expression is now:

$$2x + 2y - 5x$$

Combine the x terms: $2x - 5x = -3x$.

$2(x + y) - 5x$ simplifies to $-3x + 2y$.

7. Use the distributive law to find $w(v - 1)$. Multiply w by v and multiply w by -1:

$$w(v - 1) = wv - w$$

The expression is now:

$$wv + wv - w$$

Combine the wv terms: $wv + wv = 2wv$.

$wv + w(v - 1)$ simplifies to $2wv - w$.

8. This expression has two f^5g terms, one f^2 term, and two g terms.

Combine the f^5g terms: $6f^5g - 3f^5g = 3f^5g$.

Combine the g terms: $-4g + g = -3g$.

$6f^5g + 3f^2 - 4g - 3f^5g + g$ simplifies to $3f^5g + 3f^2 - 3g$.

9. This expression has one a^2 term, two b terms, and one constant term before we multiply.

Use the distributive law to find $5(2b + 3)$. Multiply 5 by $2b$ and multiply 5 by 3:

$$5(2b + 3) = 10b + 15$$

The expression is now:

$$-9a^2 + 10b + 15 - b$$

Combine the b terms: $10b - b = 9b$.

$-9a^2 + 5(2b + 3) - b$ simplifies to $-9a^2 + 9b + 15$.

10. This expression has three k terms and three m terms before we multiply.

Use the distributive law to find $4(k + m)$. Multiply 4 by k and multiply 4 by m:

$$4(k + m) = 4k + 4m$$

Use the distributive law again to find $-9(3k - 2m)$. Multiply -9 by $3k$ and multiply -9 by $-2m$:

$$-9(3k - 2m) = -27k + 18m$$

The expression is now:

$$4k + 4m - 27k + 18m + k + m$$

Combine the k terms: $4k - 27k + k = -22k$.

Combine the m terms: $4m + 18m + m = 23m$.

$4(k + m) - 9(3k - 2m) + k + m$ simplifies to $-22k + 23m$.

Practice 2

1. The terms $3r^2$ and $4t$ are unlike, so this expression cannot be simplified.
   Substitute 3 for r and -2 for t:
   $$3(3)^2 + 4(-2) = 3(9) - 8 = 27 - 8 = 19$$

2. This expression has two r terms and two t terms.
   Combine the r terms: $-r + 9r = 8r$.
   Combine the t terms: $5t - 2t = 3t$.
   The expression is now $8r + 3t$.
   Substitute 3 for r and -2 for t:
   $$8(3) + 3(-2) = 24 - 6 = 18$$

3. This expression has two r terms, one t term, and one t^2 term.
   Combine the r terms: $6r - 2r = 4r$.
   The expression is now $4r - 3t^2 + 4t$.
   Substitute 3 for r and -2 for t:
   $$4(3) - 3(-2)^2 + 4(-2) = 4(3) - 3(4) - 8 = 12 - 12 - 8 = -8$$

4. Use the distributive law to find $-2r(8 + t)$. Multiply $-2r$ by 8 and multiply $-2r$ by t:
   $$-2r(8 + t) = -16r + -2rt$$
   This expression, $4rt - 16r - 2rt$, has one r term and two rt terms.
   Combine the rt terms: $4rt - 2rt = 2rt$.
   The expression is now $2rt - 16r$.
   Substitute 3 for r and -2 for t:
   $$2(3)(-2) - 16(3) = 6(-2) - 48 = -12 - 48 = -60$$

5. This expression has two a terms, two b terms, and one c term.
   Combine the a terms: $2a - 5a = -3a$.
   Combine the b terms: $3b - 4b = -b$.
   The expression is now $-3a - b - 4c$.
   Substitute -3 for a, 6 for b, and 1 for c:
   $$-3(-3) - (6) - 4(1) = 9 - 6 - 4 = 3 - 4 = -1$$

6. Use the distributive law to find $4(a - b + 2c)$. Multiply each term in parentheses by 4:

$$4(a - b + 2c) = 4a - 4b + 8c$$

Use the distributive law again to find $c(a - 3)$. Multiply both terms in parentheses by c:

$$c(a - 3) = ac - 3c$$

This expression, $4a - 4b + 8c + ac - 3c$, has one a term, one b term, two c terms, and one ac term.

Combine the c terms: $8c - 3c = 5c$.

The expression is now $4a - 4b + 5c + ac$.

Substitute –3 for a, 6 for b, and 1 for c:

$$4(-3) - 4(6) + 5(1) + (-3)(1) = -12 - 24 + 5 - 3 = -36 + 5 - 3 = -31 - 3 = -34$$

7. Use the distributive law to find $b(3a + c)$. Multiply each term in parentheses by b:

$$b(3a + c) = 3ab + bc$$

This expression, $3c^2 - 3ab + 3ab + bc$, has one c^2 term, two ab terms, and one bc term.

Combine the ab terms: $-3ab + 3ab = 0$.

The expression is now $3c^2 + bc$.

Substitute 6 for b and 1 for c:

$$3(1)^2 + (6)(1) = 3(1) + (6)(1) = 3 + 6 = 9$$

8. Use the distributive law to find $2b(c - a)$. Multiply each term in parentheses by $2b$:

$$2b(c - a) = 2bc - 2ab$$

This expression, $5b + 2bc - 2ab + 3ab$, has one b term, one bc term, and two ab terms.

Combine the ab terms: $-2ab + 3ab = ab$.

The expression is now $5b + 2bc + ab$.

Substitute –3 for a, 6 for b, and 1 for c:

$$5(6) + 2(6)(1) + (-3)(6) = 30 + 12(1) - 18 = 30 + 12 - 18 = 42 - 18 = 24$$

9. Use the distributive law to find $6(3w + x)$. Multiply both terms in parentheses by 6:

$$6(3w + x) = 18w + 6x$$

Use the distributive law again to find $-3(w - z)$. Multiply both terms in parentheses by –3:

$$-3(w - z) = -3w + 3z$$

This expression, $18w + 6x + 2y - 3w + 3z$, has two w terms, one x term, one y term, and one z term.

Combine the w terms: $18w - 3w = 15w$.

The expression is now $15w + 6x + 2y + 3z$.

Substitute 1 for w, -2 for x, 3 for y, and -4 for z:

$$15(1) + 6(-2) + 2(3) + 3(-4) = 15 - 12 + 6 - 12 = 3 + 6 - 12 = 9 - 12 = -3$$

10. Use the distributive law to find $5(z + 4)$. Multiply both terms in parentheses by 5:

$$5(z + 4) = 5z + 20$$

Use the distributive law again to find $-10(\frac{1}{2}z + y)$. Multiply both terms in parentheses by -10:

$$-10((\tfrac{1}{2}z + y) = -5z - 10y$$

This expression, $w^2 - 3x + 5z + 20 - 5z - 10y$, has one w^2 term, one x term, one y term, two z terms, and one constant.

Combine the z terms: $5z - 5z = 0$.

The expression is now $w^2 - 3x + 20 - 10y$.

Substitute 1 for w, -2 for x, and 3 for y:

$$(1)^2 - 3(-2) + 20 - 10(3) = 1 + 6 + 20 - 30 = 7 + 20 - 30 = 27 - 30 = -3$$

algebra word expressions

We are mere operatives, empirics, and egotists until
we learn to think in letters instead of figures.
—OLIVER WENDELL HOLMES (1841–1935)
AMERICAN PHYSICIAN AND WRITER

In this lesson, you'll learn how to translate word expressions into numerical and algebraic expressions.

YOU MIGHT NOT believe it, but real-life situations can be turned into algebra. In fact, many real-life situations that involve numbers are really algebra problems. Let's say it costs \$5 to enter a carnival, and raffle tickets cost \$2 each. The number of tickets bought by each person varies, so we can use a variable, x, to represent that number. The total spent by a person is equal to $2x + 5$. If we know how many raffle tickets a person bought, we can evaluate the expression to find how much money that person spent in total.

How do we turn words into algebraic expressions? First, let's look at numerical expressions. Think about how you would describe "2 + 3" in words. You might say, *Two plus three, two added to three, two increased by three, the sum of two and three*, or *the total of two and three*. You might even think of other ways to describe 2 + 3. Each of these phrases contains a keyword that signals addition. In the first phrase, the word *plus* tells us that the numbers should be added. In the last phrase, the word *total* tells us that we will be adding.

For each of the four basic operations, there are keywords that can signal which operation will be used. The following chart lists some of those keywords and phrases.

Addition	Subtraction	Multiplication	Division
combine	take away	times	share
together	difference	product	quotient
total	minus	factors	percent
plus	decrease	each	out of
sum	left	every	average
altogether	less than	increase	each
and	fewer	of	per
increase		double, triple, . . .	

Some words and phrases can signal more than one operation. For example, the word *each* might mean multiplication, as it did in the raffle ticket example at the beginning of the lesson. However, if we were told that a class reads 250 books and we are looking for how many books each student read, *each* would signal division.

Let's start with some basic phrases. *The total of three and four* would be an addition expression: 3 + 4. The keyword *total* tells us to add three and four. The phrase *the difference between ten and six* is a subtraction phrase, because of the keyword *difference*. The difference between ten and six is 10 − 6.

Sometimes, the numbers given in a phrase appear in the opposite order that we will use them when we form our number sentence. This can happen with subtraction and division phrases, where the order of the numbers is important. The order of the addends in an addition sentence, or the order of the factors in a multiplication sentence, does not matter. The phrase *one fewer than five* is a subtraction phrase, but be careful. *One fewer than five* is 5 − 1, not 1 − 5.

Some phrases combine more than one operation: *ten more than five minus three* can be written as 10 + 5 − 3 (or 5 − 3 + 10). As we learned in Lesson 4, the order of operations is important, and it is just as important when forming number sentences from phrases. *Seven less than eight times negative two* is (8)(−2) − 7. We must show that 8 and −2 are multiplied, and that 7 is subtracted from that product.

Although multiplication comes before subtraction in the order of operations, a phrase might be written in such a way that subtraction must be performed first. The phrase *five times the difference between eleven and four* means that

4 must be subtracted from 11 before multiplication occurs. We must use parentheses to show that subtraction should be performed before multiplication: $5(11 - 4)$.

....

TIP: If you are writing a phrase as a number sentence, and you know that one operation must be performed before another, place that operation in parentheses. Even if the parentheses are unnecessary, it is never wrong to place parentheses around the operation that is to be performed first.

....

Practice 1

Write each phrase as a numerical expression.

1. *five increased by two*

2. *the quotient of sixteen and four*

3. *four less than the sum of two and nine*

4. *negative twelve times three fewer than fifteen*

5. *the difference between the total of one and seven and the product of six and three*

WRITING PHRASES AS ALGEBRAIC EXPRESSIONS

Writing algebraic phrases as algebraic expressions is very similar to writing numerical phrases as numerical expressions. What is the difference between numerical phrases and algebraic phrases? Numerical phrases, like the ones we have seen so far in this lesson, contain only numbers whose values are given (such as *five* or *ten*). Algebraic phrases contain at least one unknown quantity. That unknown is usually referred to as *a number*, as in "five times a number." A second unknown is usually referred to as *another number* or *a second number*.

The variable x is usually used to represent "a number" in these expressions, although we could use any letter. Just as the phrase *ten more than five* is written as $10 + 5$, the phrase *ten more than a number* is written as $10 + x$.

When a phrase refers to two unknown values, we usually use x to represent the first number and y to represent the second number. *Twice a number plus four times another number* would be written as $2x + 4y$.

Practice 2

Write each phrase as an algebraic expression. Use x to represent one number and y to represent a second number.

1. *negative nine times a number*

2. *a number combined with six*

3. *eight take away a number*

4. *a number shared by ten*

5. *two plus one-half of a number*

6. *the total of a number and negative one times another number*

7. *the quotient of seven and a number minus a second number*

8. *eighteen less than a number multiplied by another number*

9. *the difference between the square of a number and ten times another number*

10. *the total of six times a number and the difference between another number and nine*

ANSWERS

Practice 1

1. The keyword *increased* signals addition. *Five increased by two* is $5 + 2$.
2. The keyword *quotient* signals division. *The quotient of sixteen and four* is $16 \div 4$.

3. The keywords *less than* signal subtraction and the keyword *sum* signals addition. *The sum of two and nine* is $2 + 9$, and *four less than* that sum is $(2 + 9) - 4$.

4. The keywords *fewer than* signal subtraction and the keyword *times* signals multiplication. *Three fewer than fifteen* is $15 - 3$. *Negative twelve times* that quantity is $-12(15 - 3)$.

5. The keyword *difference* signals subtraction, but before we can subtract, we have to translate *the total of one and seven* and *the product of six and three* into numbers and operations. The keyword *total* signals addition. *The total of one and seven* is $(1 + 7)$. The keyword *product* signals multiplication. *The product of six and three* is $(6)(3)$. Now we are ready to write the difference between the two quantities. *The difference between the total of one and seven and the product of six and three* is $(1 + 7) - (6)(3)$.

Practice 2

1. The keyword *times* signals multiplication. *Negative nine times a number* is $-9x$.

2. The keyword *combined* signals addition. *A number combined with six* is $x + 6$.

3. The keyword phrase *take away* signals subtraction. *Eight take away a number* is $8 - x$.

4. The keyword *shared* signals division. *A number shared by ten* is $\frac{x}{10}$.

5. The keyword *plus* signals addition. The keyword *of* signals multiplication. *One-half of a number* is $\frac{1}{2}$ times a number, or that number divided by two. *Two plus one-half of a number* is $2 + \frac{x}{10}$.

6. The keyword *total* signals addition and the keyword *times* signals multiplication. *The total of a number and negative one times another number* is $x + -1y$, which simplifies to $x - y$.

7. The keyword *quotient* signals division and the keyword *minus* signals subtraction. *The quotient of seven and a number minus a second number* is $\frac{7}{x - y}$.

8. The keyword phrase *less than* signals subtraction and the keyword *multiplied* signals multiplication. A number multiplied by another number is xy. *Eighteen less than a number multiplied by another number* is $xy - 18$.

9. The keyword *difference* signals subtraction and the keyword *times* signals multiplication. The square of a number is a number times itself, or x^2.

Ten times another number is $10y$. *The difference between the square of a number and ten times another number* is $x^2 - 10y$.

10. The keyword *total* signals addition, the keyword *times* signals multiplication, and the keyword *difference* signals subtraction. *The difference between another number and nine* is $y - 9$. *The total of six times a number and that difference* is $6x + (y - 9)$.

factoring

Everything should be made as simple as possible, but not simpler.
—ALBERT EINSTEIN (1879–1955)
GERMAN-AMERICAN THEORETICAL PHYSICIST

In this lesson, you'll learn how to factor single-variable and multivariable algebraic expressions.

AS WE LEARNED earlier, two numbers or variables that are multiplied are called factors. We multiply factors to find a product. When we break down a product into its factors, that's called **factoring**.

To factor a whole number, such as 6, we list every other whole number that divides evenly into 6. The numbers 1, 2, 3, and 6 all divide evenly into 6. Every number has itself and 1 as factors, because 1 multiplied by any number is that number. Numbers whose only whole factors are 1 and itself are called **prime numbers**. Numbers that have at least one other factor are called **composite numbers**.

....

TIP: Every whole number is either prime or composite except the number 1, which is considered neither prime nor composite.

....

PRIME FACTORIZATION

The **prime factorization** of a number is a multiplication sentence made up of only prime numbers, the product of which is the number. For example, the prime factorization of 6 is (2)(3), because 2 and 3 are both prime numbers, and they multiply to 6.

The number 7 is a prime number; its only factors are 1 and 7. Every variable has exactly two factors: 1 and itself. The factors of x are 1 and x. If a term has a coefficient, some of the factors of that term are its coefficient and its variables (with their exponents). That coefficient can be broken down into its factors, and the variables with their exponents can be broken down into their factors.

The term $8x$ can be factored into $(8)(x)$, because $8x$ means "8 multiplied by x." 8 can be factored into (2)(4), and 4 can be factored into (2)(2). The factorization of $8x$ is $(2)(2)(2)(x)$. None of the values in parentheses can be factored any further.

The term $3x^2$ can be factored into $(3)(x^2)$. 3 is a prime number, so it cannot be factored any further, but x^2 is equal to $(x)(x)$. $3x^2 = (3)(x)(x)$.

FACTORING AN EXPRESSION

Now that we know how to factor a single term, let's look at how to factor an expression. $(5x + 10)$ is made up of two terms, $5x$ and 10. When we are breaking down an expression of two or more terms, we say that we are looking to "factor out" any numbers or variables that divide both terms evenly. 10 is divisible by 2, but $5x$ is not, so we cannot factor out 2. $5x$ is divisible by x, because $5x = (5)(x)$, but 10 is not divisible by x. 10 and $5x$ are both divisible by 5, however. We can divide each term by 5: $5x \div 5 = x$ and $10 \div 5 = 2$. We write 5 outside the parentheses to show that both terms inside the parentheses must be multiplied by 5: $5(x + 2)$.

TIP: To check that you've factored correctly, multiply your factors. You should get back your original expression.

Only one of the two terms in $5x + 10$ contained the variable x, so we were not able to factor out x. If every term in an expression contains the same variable, you can factor out that variable. Sometimes, you can factor out more than one of the same variable.

Every term in the expression $(x^5 + x^3)$ has x in it. The first term, x^5, can be factored into $(x)(x)(x)(x)(x)$. The second term, x^3, can be factored into $(x)(x)(x)$. Both terms have x as a common factor. In fact, both terms have three x's as common factors. We can factor out three x's, or x^3, from each term: $(x^5 + x^3) = x^3(x^2 + 1)$.

Let's look closely at how each term was factored. When we factored x^3 out of x^5, we were left with x^2. Remember: the exponent of a base is the number of times that base is multiplied by itself. Every time we factor an x out of x^5, its exponent decreases by 1. Because we factored three x's out of x^5, we are left with two x's, or x^2.

The second term in $(x^5 + x^3)$ is x^3. Whenever we factor out an entire term, the number 1 is left behind in its place, because 1 times any quantity is that quantity. When we factor a variable out of an expression, the exponent of the factor is equal to the lowest exponent that variable has in the expression. That might sound confusing, so let's look at a few examples:

- We can factor x^5 out of $(x^{10} + 3x^6 + 2x^5)$, because every term has an x in it, and the lowest value of an exponent of x is 5, in the last term. $(x^{10} + 3x^6 + 2x^5)$ factors into $x^5(x^5 + 3x + 2)$.
- We can factor a out of $(10a^8 - 3a + 7a^3)$, because every term has an a in it, and the lowest value of an exponent of a is 1, in the middle term. $(10a^8 - 3a + 7a^3)$ factors into $a(10a^7 - 3 + 7a^2)$.
- We can factor g^9 out of $(g^9 - 4g^{10} + 11g^{13})$, because every term has a g in it, and the lowest value of an exponent of g is 9, in the first term. $(g^9 - 4g^{10} + 11g^{13})$ factors into $g^9(1 - 4g + 11g^4)$.

If you factor a variable out of every term, the term with the lowest exponent for that variable will no longer contain that variable. In $(x^5 + x^3)$, the second term contained the lowest exponent of x, and when we had finished factoring, the second term no longer contained the variable x: $x^2(x^3 + 1)$. In $(g^9 - 4g^{10} + 11g^{13})$, the first term contained the lowest exponent of g, and when we had finished factoring, the first term no longer contained the variable g: $g^9(1 - 4g + 11g^4)$.

Example

Factor $18x^3 + 12x$.

First, find the largest whole number that divides evenly into both coefficients. The factors of 18 are 1, 2, 3, 6, 9, and 18. The factors of 12 are 1, 2, 3, 4, 6, and 12. The greatest common factor is 6. We can factor out a 6 from both terms. Next, look at the variables. Both terms have x in common. The lowest exponent of x is 1, in the second term. We can factor out x from both terms.

To factor $6x$ out of both terms, divide both terms by $6x$: $18x^3 \div 6x = 3x^2$ and $12x \div 6x = 2$. $18x^3 + 12x = 6x(3x^2 + 2)$. We can check our work by multiplying: $6x(3x^2 + 2) = (6x)(3x^2) + (6x)(2) = 18x^3 + 12x$, so we have factored correctly.

Some expressions can only have a whole number factored out, and other expressions can only have a variable factored out. Some expressions, such as $10x + 7$, cannot be factored at all. There is no whole number (other than 1) that divides evenly into 10 and 7, and there is no variable common to both terms.

Practice 1

Factor each expression.

1. $6x + 2$

2. $5x^2 + x$

3. $15y - 20y^4$

4. $-2x^5 + 4x^4$

5. $13h^2 - 13h$

6. $a^4 - a^3 + a^2$

7. $6s + 15s^4 - 21s^8$

8. $14z^6 - 35z^9 - 42z^4$

9. $9p^3 + 27p^5 - 6p^{10} + 12p^{12}$

10. $-20q^8 - 8q^{20} - 40q^{12} - 80q^{10}$

FACTORING MULTIVARIABLE EXPRESSIONS

Just as with single-variable expressions, a coefficient or variable can only be factored out of a multivariable expression if the coefficient or variable appears in every term. If two variables are common to every term, then two variables can be factored out of the expression.

Example

Factor $9x^5y^2 + 12xy^3$.

Start with the coefficients. The greatest common factor of 9 and 12 is 3, so 3 can be factored out of the expression.
Next, check to see if each variable exists in each term: x is in both terms and so is y. Both can be factored out of the expression. The smallest exponent of x is 1 (in $12xy^3$) and the smallest exponent of y is 2 (in $9x^5y^2$). Divide each term by $3xy^2$: $9x5y^2 \div 3xy^2 = 3x^4$, and $12xy^3 \div 3xy^2 = 4y$.
$$9x^5y^2 + 12xy^3 = 3xy^2(3x^4 + 4y)$$

Example

Factor $7x^6y^6 - 2x^5 + 8x^5y^4$.

The greatest common factor of 7, 2, and 8 is 1, so no constant can be factored out of the expression. The variable x is common to every term, so it can be factored out, but the variable y is not in the middle term, $-2x^5$, so it cannot be factored out. The smallest exponent of x is 5, in both the second and third terms, so we can factor x^5 out of every term.
Divide each term by x^5: $7x^6y^6 \div x^5 = 7xy^6$, $2x^5 \div x^5 = 2$, and $8x^5y^4 \div x^5 = 8y^4$.
$$7x^6y^6 - 2x^5 + 8x^5y^4 = x^5(7xy^6 - 2 + 8y^4)$$

Example

Factor $16a^4b^9c^4 - 24a^3b^{12}c^6 + 40a^5b^7c^5$.

The greatest common factor of the coefficients, 16, 24, and 40, is 8. The variables a, b, and c are common to every term. The smallest exponent of a is 3, the smallest exponent of b is 7, and the smallest exponent of c is 4, so $8a^3b^7c^4$ can be factored out of every term: $16a^4b^9c^4 \div 8a^3b^7c^4 = 2ab^2$, $24a^3b^{12}c^6 \div 8a^3b^7c^4 = 3b^5c^2$, and $40a^5b^7c^5 \div 8a^3b^7c^4 = 5a^2c$.
$$16a^4b^9c^4 - 24a^3b^{12}c^6 + 40a^5b^7c^5 = 8a^3b^7c^4(2ab^2 - 3b^5c^2 + 5a^2c)$$

Practice 2

Factor each expression.

1. $18g^8h^2 - 45g^2h^8$

2. $-36x^7y^3 + 12y^2$

3. $5a^9bc^3 - 11a^6b^5c^5$

4. $16j^{11}k^3l^2 + 48j^7k^{10}l^4 - 28j^9k^6l^6$

5. $50r^5t^4 + 35s^5t^4 + 45r^5s^5t^4$

ANSWERS

Practice 1

1. The factors of $6x$ are x, 1, 2, 3, and 6, and the factors of 2 are 1 and 2. Both numbers have 2 as a common factor, so we can factor a 2 out of each term. Only the first term has an x, so we cannot factor out an x.
Divide both terms by 2: $6x \div 2 = 3x$, and $2 \div 2 = 1$.
Write 2 outside the parentheses to show that both terms are multiplied by it: $2(3x + 1)$.

2. The factors of $5x^2$ are x^2, 1, and 5, and the factors of x are 1 and x. Both terms have x as a common factor, so we can factor an x out of each term.
Divide both terms by $5x^2 \div x = 5x$, and $x \div x = 1$.
Write x outside the parentheses to show that both terms are multiplied by it: $x(5x + 1)$.

3. The factors of $15y$ are y, 1, 3, 5, and 15, and the factors of $20y^4$ are y^4, 1, 2, 4, 5, 10, and 20. Both terms have 5 and y as common factors, so we can factor $5y$ out of each term.
Divide both terms by $5y$: $15y \div 5y = 3$, and $20y^4 \div 5y = 4y^3$.
Write $5y$ outside the parentheses to show that both terms are multiplied by it: $5y(3 - 4y^3)$.

4. The factors of $-2x^5$ are -1, 2, x^5, and the factors of $4x^4$ are 1, 2, 4, and x^4. Both terms have 2 and x as common factors. In fact, we can factor $2x^4$ out of both terms.
Divide both terms by $2x^4$: $-2x^5 \div 2x^4 = -x$, and $4x^4 \div 2x^4 = 2$.
$-2x^5 + 4x^4$ factors into $2x^4(-x + 2)$.

5. The factors of $13h^2$ are h^2, 1, and 13, and the factors of $13h$ are h, 1, and 13. Both terms have 13 and h as common factors.
Divide both terms by $13h$: $13h^2 \div 13h = h$, and $13h \div 13h = 1$.
$13h^2 - 13h$ factors into $13h(h - 1)$.

6. Each term in $a^4 - a^3 + a^2$ has no coefficient, but a is a common factor of every term. The smallest exponent of a is 2, in the term a^2.
   Divide every term by a^2: $a^4 \div a^2 = a^2$, $a^3 \div a^2 = a$, and $a^2 \div a^2 = 1$.
   $a^4 - a^3 + a^2$ factors into $a^2(a^2 - a + 1)$.

7. The greatest common factor of 6, 15, and 21 is 3. Every term has s as a common factor, and the smallest exponent of s is 1, in the term $6s$.
   Divide every term by $3s$: $6s \div 3s = 2$, $15s^4 \div 3s = 5s^3$, and $21s^8 \div 3s = 7s^7$.
   $6s + 15s^4 - 21s^8$ factors into $3s(2 + 5s^3 - 7s^7)$.

8. The greatest common factor of 14, 35, and 42 is 7. Every term has z as a common factor, and the smallest exponent of z is 4, in the term $42z^4$.
   Divide every term by $7z^4$: $14z^6 \div 7z^4 = 2z^2$, $35z^9 \div 7z^4 = 5z^5$, and $42z^4 \div 7z^4 = 6$.
   $14z^6 - 35z^9 - 42z^4$ factors into $7z^4(2z^2 - 5z^5 - 6)$.

9. The greatest common factor of 9, 27, 6, and 12 is 3. Every term has p as a common factor, and the smallest exponent of p is 3, in the term $9p^3$.
   Divide every term by $3p^3$: $3p^3 \div 3p^3 = 3$, $27p^5 \div 3p^3 = 9p^3$, $6p^{10} \div 3p^3 = 2p^7$, and $12p^{12} \div 3p^3 = 4p^9$.
   $9p^3 + 27p^5 - 6p^{10} + 12p^{12}$ factors into $3p^3(3 + 9p^2 - 2p^7 + 4p^9)$.

10. The greatest common factor of 20, 8, 40, and 80 is 4. Since every term is negative, -1 should be factored out of the expression, too. Every term has p as a common factor, and the smallest exponent of q is 8, in the term $-20q^8$.
    Divide every term by $-4q^8$: $-20q^8 \div -4q^8 = 5$, $-8q^{20} \div -4q^8 = 2q^{12}$, $-40q^{12} \div -4q^8 = 10q^4$, and $-80q^{10} \div -4q^8 = 20q^{10}$.
    $-20q^8 - 8q^{20} - 40q^{12} - 80q^{10}$ factors into $-4q^8(5 + 2q^{12} + 10q^4 + 20q^2)$.

Practice 2

1. The greatest common factor of 18 and 45 is 9. The variables g and h are common to both terms. The smallest exponent of g is 2 and the smallest exponent of h is 2, so $9g^2h^2$ can be factored out of each term.
   $$18g^8h^2 \div 9g^2h^2 = 2g^6, \text{ and } 45g^2h^8 \div 9g^2h^2 = 5h^6$$
   $$18g^8h^2 - 45g^2h^8 = 9g^2h^2(2g^6 - 5h^6)$$

2. The greatest common factor of 36 and 12 is 12. The variable y is common to both terms, but the variable x is not. The smallest exponent of y is 2, so $12y^2$ can be factored out of each term.
   $$-36x^7y^3 \div 12y^2 = -3x^7y, \text{ and } 12y^2 \div 12y^2 = 1$$
   $$-36x^7y^3 + 12y^2 = 12y^2(-3x^7y + 1)$$

3. The coefficients, 5 and 11, are both prime. No constant can be factored out of the expression. The variables a, b, and c are common to both terms. The smallest exponent of a is 6, the smallest exponent of b is 1, and the smallest exponent of c is 3: a^6bc^3 can be factored out of each term.

$$5a^9bc^3 \div a^6b^5c^5 = 5a^3, \text{ and } 11a^6b^5c^5 \div a^6bc^3 = 11b^4c^2$$
$$5a^9bc^3 - 11a^6b^5c^5 = a^6bc^3(5a^3 - 11b^4c^2)$$

4. The greatest common factor of 16, 48, and 28 is 4. The variables j, k, and l are common to each term. The smallest exponent of j is 7, the smallest exponent of k is 3, and the smallest exponent of l is 2: $4j^7k^3l^2$ can be factored out of each term.

$$16j^{11}k^3l^2 \div 4j^7k^3l^2 = 4j^4, 48j^7k^{10}l^4 \div 4j^7k^3l^2 = 12k^7l^2, \text{ and } 28j^9k^6l^6 \div 4j^7k^3l^2 = 7j^2k^3l^4$$
$$16j^{11}k^3l^2 + 48j^7k^{10}l^4 - 28j^9k^6l^6 = 4j^7k^3l^2(4j^4 + 12k^7l^2 - 7j^2k^3l^4)$$

5. The greatest common factor of 50, 35, and 45 is 5. The variable t is common to each term, but the variables r and s are not in every term. The exponent of t is 4 in every term, so $5t^4$ can be factored out of every term.

$$50r^5t^4 \div 5t^4 = 10r^5, 35s^5t^4 \div 5t^4 = 7s^5, \text{ and } 45r^5s^5t^4 \div 5t^4 = 9r^5s^5.$$
$$50r^5t^4 + 35s^5t^4 + 45r^5s^5t^4 = 5t^4(10r^5 + 7s^5 + 9r^5s^5)$$

exponents

I was x years old in the year x^2.
—AUGUSTUS DE MORGAN (1806–1871)
BRITISH MATHEMATICIAN

In this lesson, you'll learn how to work with terms that have positive and negative exponents, and we'll introduce fractional exponents.

IN LESSON 2, you learned that you can only add and subtract terms that have like bases and exponents. In Lesson 3, you learned how to multiply terms. After multiplying the coefficients, if two terms have the same base, you added their exponents. In that same lesson, you learned that when dividing terms, if the terms had the same base, you subtract the exponent of the divisor from the exponent of the dividend. What else can we do with exponents? For starters, we can raise a base with an exponent to another exponent.

You already know that x^4 means $(x)(x)(x)(x)$. But what does $(x^4)^2$ mean? The term x^4 is raised to the second power. When a term is raised to the second power, or squared, we multiply it by itself: $(x^4)^2 = (x^4)(x^4)$. Now we have a multiplication problem. Keep the base and add the exponents: $4 + 4 = 8$, so $(x^4)^2 = (x^4)(x^4) = x^8$.

What does $(x^3)^6$ mean? It means multiply x^3 six times: $(x^3)(x^3)(x^3)(x^3)(x^3)(x^3)$. Adding the exponents, we can see that $(x^3)^6 = x^{18}$. Is there an easier way to find $(x^3)^6$, without having to write out a long multiplication sentence? Yes. When raising a base with an exponent to another exponent, multiply the exponents.

Look again at $(x^3)^6$: $(3)(6) = 18$, and $(x^3)^6 = x^{18}$.

Example

What is $(x^5)^8$?

Multiply the exponents: $(5)(8) = 40$, so $(x^5)^8 = x^{40}$.

If the term has a coefficient, raise the coefficient to the exponent, too. To find $(4k^2)^7$, find $(4^7)(k^2)^7$: $(4^7) = 16,384$. $(2)(7) = 14$, so $(k^2)^7 = k^{14}$. $(4k^2)^7 = 16,384k^{14}$.

Be sure to look carefully at a term before multiplying exponents. In the expression $(5x^5)^8$, the entire term, coefficient and base, is raised to the eighth power. In the expression $6(a^4)^5$, only the base and its exponent, a^4, is raised to the fifth power. Why? Remember the order of operations. In $(5x^5)^8$, $5x^5$ is in parentheses, which means do multiplication in parentheses before the term is raised to the eighth power. Because we cannot simplify $5x^5$ any further, raise the entire term to the eighth power. In the expression $6(a^4)^5$, the 6 is outside the parentheses. Exponents come before multiplication in the order of operations, so raise a^4 to the fifth power before multiplying by 6.

Example

What is $3(2m^2)^5$?

The term $2m^2$ is raised to the fifth power and multiplied by 3. Because exponents come before multiplication in the order of operations, begin by raising $2m^2$ to the fifth power: $2^5 = 32$ and $(m^2)^5 = m^{10}$, so $(2m^2)^5 = 32m^{10}$. Now, multiply by 3: $3(32m^{10}) = 96m^{10}$.

Practice 1

Solve:

1. $(b^2)^3$

2. $(n^6)^6$

3. $(3v^7)^4$

4. $4(r^5)^{10}$

5. $5(3w^8)^2$

If a term contains more than one variable raised to an exponent, each variable must be raised to the exponent. To find $(a^2b^2)^6$, raise both a^2 and b^2 to the sixth power. $(a^2)^6 = a^{12}$ and $(b^2)^6 = b^{12}$, so $(a^2b^2)^6 = a^{12}b^{12}$.

Example
What is $12(f^3g^2h^7)^4$?

Raise each part within the parentheses to the fourth power: $(f^3)^4 = f^{12}$, $(g^2)^4 = g^8$, and $(h^7)^4 = h^{28}$. $(f^3g^2h^7)^4 = f^{12}g^8h^{28}$.
Now, multiply by 12: $(12)(f^{12}g^8h^{28}) = 12f^{12}g^8h^{28}$.

NEGATIVE EXPONENTS

So far, all of the exponents we have seen in this lesson have been positive. But we can work with negative exponents, too. Remember: a positive number multiplied by a negative number will give you a negative product, so a positive exponent multiplied by a negative exponent will give you a negative exponent: $(x^3)^{-5} = x^{-15}$, because $(3)(-5) = -15$.

Example
What is $(d^{-3})^{-9}$?

A negative number multiplied by a negative number will give you a positive product: $(-3)(-9) = 27$, so $(d^{-3})^{-9} = d^{27}$.

In Lesson 3, we learned that bases with negative exponents appear as positive exponents in the denominator when the term is expressed as a fraction. When a coefficient is raised to a negative exponent, express the result as a fraction. Make the numerator of the fraction 1, and place the base in the denominator with a positive exponent: 4^{-2} is equal to $\frac{1}{4^2}$, which is $\frac{1}{16}$.

Example
What is $(5m^3)^{-4}$?

$5m^3$ is in parentheses, so the entire term is raised to the negative fourth power: $(3)(-4) = -12$, so $(m^3)^{-4} = m^{-12}$. $(5)^{-4} = \frac{1}{5^4}$, which is $\frac{1}{625}$.
$(5m^3)^{-4} = \frac{1}{625}m^{-12}$, which could also be written as $\frac{1}{625m^{12}}$.

ZERO EXPONENTS

An expression with a term that has an exponent of 0 is the easiest to simplify. Any value—number or variable—raised to the power of 0 is 1: $1^0 = 1$, $2^0 = 2$, $100^0 = 1$, $x^0 = 1$, and $(abcd)^0 = 1$. We must still remember our order of operations though. The term $3x^0$ has only x raised to the power of zero, so this term is equal to $3(x^0) = 3(1) = 3$. Only the base that is raised to the zero power is equal to 1. If a term has a coefficient or another base that is not raised to the zero power, those parts of the term must still be evaluated.

> **Example**
> What is $4a^0b^2$?
>
> $a^0 = 1$, so $4(a^0)(b^2) = 4(1)(b^2) = 4b^2$

Practice 2

1. $(pqr)^9$

2. $(3j^6k^2)^3$

3. $-10(2s^8t^5)^5$

4. $(u^7)^{-6}$

5. $5(e^{-5})^{12}$

6. $(6a^{-11})^3$

7. $(2z^{-2})^{-2}$

8. $(g^3h^{-4})^{-8}$

9. $(4p^3q^8)^0$

10. $2(k^{13}m^0n-1)^{-1}$

FRACTIONAL EXPONENTS

A constant or variable can also be raised to a fractional exponent. The numerator of the exponent is the power to which the constant or variable is raised. The denominator of the exponent is the root that must be taken of the constant or variable. For instance $x^{\frac{3}{2}}$, means that x is raised to the third power, and then the second, or square, root of x must be taken. The root symbol, or radical symbol, looks like this: $\sqrt{}$. The square root of x is $\sqrt{x}$. To show that x is raised to the third power, we put the exponent inside the radical: $\sqrt{x^3}$, therefore, $x^{\frac{3}{2}} = \sqrt{x^3}$.

If we needed to find a larger root of x, such as 5 or 6, that number would be placed just outside the radical symbol: $x^{\frac{2}{5}}$ is the fifth root of x raised to the second power. We place x^2 inside the radical and 5 outside the radical: $x^{\frac{2}{5}} = \sqrt[5]{x^2}$. We will learn more about radicals and how to evaluate them in Lesson 11.

ANSWERS

Practice 1

1. b^2 is raised to the third power. Multiply the exponents: $(2)(3) = 6$, and $(b^2)^3 = b^6$.

2. n^6 is raised to the sixth power. Multiply the exponents: $(6)(6) = 36$, and $(n^6)^6 = n^{36}$.

3. $3v^7$ is raised to the fourth power.
   Raise 3 to the fourth power and raise v^7 to the fourth power: $3^4 = 81$.
   To find $(v^7)^4$, multiply the exponents: $(7)(4) = 28$, $(v^7)^4 = v^{28}$, and $(3v^7)^4 = 81v^{28}$.

4. r^5 is raised to the tenth power and multiplied by 4.
   Exponents come first in the order of operations, so begin by raising r^5 to the tenth power.
   Multiply the exponents: $(5)(10) = 50$, and $(r^5)^{10} = r^{50}$.
   Finally, multiply by 4: $(4)(r^{50}) = 4r^{50}$.

5. $3w^8$ is raised to the second power and multiplied by 4. Work with the exponent before multiplying.
   Raise 3 to the second power and raise w^8 to the second power. $3^2 = 9$.
   To find $(w^8)^2$, multiply the exponents: $(8)(2) = 16$, $(w^8)^2 = w^{16}$, and $(3w^8)^2 = 9w^{16}$.
   Finally, multiply $9w^{16}$ by 5: $(5)(9w^{16}) = 45w^{16}$.

Practice 2

1. pqr is raised to the ninth power. Each variable has an exponent of 1, and each exponent must be multiplied by 9. Because $(1)(9) = 9$, the exponent of each variable is 9.

   $(pqr)^9 = p^9 q^9 r^9$

2. $(3j^6 k^2)$ is raised to the third power. Each variable must have its exponent multiplied by 3, and the coefficient 3 must also be raised to the third power.

   $3^3 = 27$
   $(6)(3) = 18$, so $(j^6)^3 = j^{18}$
   $(2)(3) = 6$, so $(k^2)^3 = k^6$
   $(3j^6 k^2)^3 = 27j^{18} k^6$

3. $(2s^8 t^5)$ is raised to the fifth power, and then multiplied by -10. Each variable must have its exponent multiplied by 5, and the coefficient 2 must also be raised to the fifth power.

   $2^5 = 32$
   $(8)(5) = 40$, so $(s^8)^5 = s^{40}$
   $(5)(5) = 25$, so $(t^5)^5 = t^{25}$
   $(2s^8 t^5)^5 = 32s^{40} t^{25}$

   Finally, multiply by -10:
   $(-10)(32s^{40} t^{25}) = -320s^{40} t^{25}$

4. (u^7) is raised to the negative sixth power. Multiply 7 by -6:
   $(7)(-6) = -42$, so $(u^7)^{-6} = u^{-42}$

5. (e^{-5}) is raised to the 12th power, and then multiplied by 5. Multiply -5 by 12:
   $(-5)(12) = -60$, so $(e^{-5})^{12} = e^{-60}$

   Multiply e^{-60} by its coefficient, 5:
   $(5)(e^{-60}) = 5e^{-60}$

6. $(6a^{-11})$ is raised to the third power. The variable a must have its exponent multiplied by 3, and the coefficient 6 must also be raised to the third power.

   $6^3 = 216$
   $(-11)(3) = -33$, so $(a^{-11})^3 = a^{-33}$
   $(6a^{-11})^3 = 216a^{-33}$

7. $(2z^{-2})$ is raised to the negative second power. The variable z must have its exponent multiplied by -2, and the coefficient 2 must also be raised to the negative second power. To raise 2 to the negative second power, place the

2 with an exponent of positive 2 in the denominator of a fraction with a numerator of 1:

$$2^{-2} = \frac{1}{2^2} = \frac{1}{4}$$

$(-2)(-2) = 4$, so $(z^{-2})^{-2} = z^4$

$(2z^{-2})^{-2} = \frac{1}{4}z^4$

8. $(g^3 h^{-4})$ is raised to the negative eighth power. The variables g and h must each have their exponent multiplied by -8.

$(3)(-8) = -24$, so $(g^3)^{-8} = g^{-24}$

$(-4)(-8) = 32$, so $(h^{-4})^{-8} = h^{32}$

$(g^3 h^{-4})^{-8} = g^{-24} h^{32}$

9. The entire term in parentheses is raised to the zero power. Any quantity raised to the zero power is equal to 1.

10. $(k^{13} m^0 n^{-1})$ is raised to the negative 1 power, and then multiplied by 2. The variables k and n must each have their exponent multiplied by -1. The variable m is raised to the zero power, so it is equal to 1. Because k^{13} and n^{-1} are multiplied by $m^0 = 1$, the m term drops out of the expression.

$(13)(-1) = -13$, so $(k^{13})^{-1} = k^{-13}$

$(-1)(-1) = 1$, so $(n^{-1})^{-1} = n^1$, or n

The expression is now $2(k^{-13} n)$. Multiply by 2:

$(2)(k^{-13} n) = 2k^{-13} n$

SECTION 2

solving and graphing equations and inequalities

THE BASICS OF algebra have given us the foundation we need to start solving equations. We know how to work with expressions; an equation is just an expression with an equal sign. Some equations can be solved quickly, with just one operation, but some equations may require many operations in order to be solved.

After learning how to find the value of a variable, we will look at how to graph an equation, and how to find the equation of a line that appears on a graph. Graphs are a picture of what an equation looks like, and every point on the graph of a line represents a solution to the equation. Graphs can also help us decide if an equation is a function, or what the domain and range of the equation might be.

Next, we will see how equations are not the only relationships we can have between two quantities; inequalities can also be solved and graphed. We will learn how to solve a system of equations, and how graphs can show the solution to a system of inequalities. By the time you are through with this section, you'll be on your way to mastering algebra.

This section will teach you a variety of algebra concepts, including:

- single-step and multistep equations
- radicals and fractional exponents
- equations with radicals
- equations in slope-intercept form

- input/output tables
- the coordinate plane
- the equation of a line
- the distance between two points
- functions, domain, and range
- systems of equations
- single-variable and multivariable inequalities
- inequalities and systems of inequalities
- FOIL
- the quadratic equation and the quadratic formula

solving single-step
algebraic equations

*Strange as it may sound, the power of mathematics rests
on its evasion of all unnecessary thought and on its
wonderful saving of mental operations.*
—ERNST MACH (1838–1916)
AUSTRIAN PHYSICIST AND PHILOSOPHER

In this lesson, you'll learn how to solve one-step algebraic equations.

SO FAR, WE HAVE SEEN algebraic terms only in expressions. In this lesson, we will look at algebraic equations. An **equation** presents two expressions that are equal to each other. $3 + 6 = 9$, and even $3 = 3$ is an equation. An **algebraic equation** is an equation that includes at least one variable.

Working with algebraic expressions has taught us many of the skills we need to solve equations. The goal of solving equations is to get the variable alone on one side of the equation. If the variable is alone on one side, its value must be on the other side, which means the equation is solved.

Example
$x + 4 = 10$

This equation has one variable, x. We can see that 4 is added to x, and the sum of x and 4 is equal to 10. How do we get x alone on one side of the equation? We need to get rid of that 4. We can do that by subtracting 4 from both sides of the equation. Why do we subtract 4 from both sides? The equal sign in an equation tells us that the quantities on each side of the sign have the same value. If we perform an action on one side, such as subtraction, we must perform the same action on the other side, so that the two sides of the equation stay equal. This is the most important rule when solving equations: Whatever we do to one side of an equation, we do the same to the other side of the equation.

Let's subtract 4 from both sides:

$$x + 4 - 4 = 10 - 4$$
$$x = 6$$

Our answer is $x = 6$. This means that if we substitute 6 for x in the equation $x + 4 = 10$, the equation will remain true. Some equations have more than one answer, but this equation has just one answer.

Example
$$y - 3 = 13$$

The first step in solving an equation is to find the variable. In this equation, the variable is y: y is what we must get alone on one side of the equation.

The second step in solving an equation is determining what operation or operations are needed to get the variable alone. In the previous example, we used subtraction. Why? Because a constant, 4, was added to x. We used the opposite of addition, subtraction, to get x alone. Addition and subtraction are opposite operations. Multiplication and division are also opposite operations.

The third step is to perform the operation on both sides of the equation. If we are left with the variable alone on one side of the equation, then we have our answer. If not, then we must repeat steps two and three until we have our answer.

In the equation $y - 3 = 13$, 3 is subtracted from y. We must use the opposite of subtraction, addition, to get y alone on one side of the equation. Add 3 to both sides of the equation:

$$y - 3 + 3 = 13 + 3$$
$$y = 16$$

When a variable in an equation has a coefficient, we must use division to get the variable alone. Remember, a coefficient and a base in a term are multiplied. Division is what we use to undo the multiplication.

Example
$$6z = 60$$

Because z is multiplied by 6, divide both sides by 6 to get z alone:

$$\frac{6z}{6} = \frac{60}{6}$$
$$z = 10$$

TIP: Use the following chart to help you decide what operation to use to solve an equation.

Type of Equation	Example	Operation to Use	Solution
addition	$a + 1 = 8$	subtraction	$a + 1 - 1 = 8 - 1; a = 7$
subtraction	$b - 6 = 12$	addition	$b - 6 + 6 = 12 + 6; b = 18$
multiplication	$3c = 15$	division	$3c = 15, \frac{3c}{3} = \frac{15}{3}; c = 5$
division	$\frac{d}{4} = 8$	multiplication	$\frac{d}{4} = 8, 4(\frac{d}{4}) = 4(8); d = 32$

Sometimes, we need to perform an operation with a variable rather than a constant. The equation $2q = 4 + q$ has a variable, q, on both sides of the equation. If we subtract 4 from both sides, or divide both sides by 2, we will still have q on both sides of the equation. If we subtract q from both sides of the equation, then we will have one q, alone, on the left side of the equation:

$$2q - q = 4 + q - q$$
$$q = 4$$

Practice 1

Find the value of the variable in each equation.

1. $u + 9 = 17$

2. $-7s = 49$

3. $y - 13 = 27$

4. $\frac{f}{5} = 35$

5. $54 = 9c$

6. $h - 1 = -1$

7. $\frac{w}{8} = 6$

8. $12 = e + 19$

9. $-2p = -48$

10. $\frac{3}{4}k = 33$

11. $12 + a = 2a$

12. $9 - v = 0$

SOLVING FOR ONE VARIABLE IN TERMS OF ANOTHER

If a single equation has two variables in it, we will not be able to find the exact value of either variable. We can, however, solve for one variable in terms of the other. What does that mean? If an equation has x and y in it, we can solve for x in terms of y by getting x alone on one side of the equation. Our solution will be an expression that contains y. We could also solve for y in terms of x by getting y alone on one side of the equation.

Example
Solve for x in terms of y in $x - y = 4$.

Because this equation has two variables, we will not be able to find a constant that is equal to x or y. To solve for x in terms of y, we can add y to both sides of the equation:

$$x - y + y = 4 + y$$
$$x = 4 + y$$

This tells us that the value of x is equal to 4 greater than y. In Lessons 18 and 19, we'll learn how to solve systems of equations, which are sets of equations that contain more than one variable.

Practice 2

Solve for x in terms of y.

1. $x + 4 = y - 6$

2. $2x = 8y$

3. $x - 7 = 3y$

4. $\frac{x}{9} = y + 1$

5. $3x = -12y + 15$

ANSWERS

Practice 1

1. In the equation $u + 9 = 17$, 9 is added to u. Use the opposite operation, subtraction, to solve the equation. Subtract 9 from both sides of the equation:
$$u + 9 - 9 = 17 - 9$$
$$u = 8$$

2. In the equation $-7s = 49$, s is multiplied by -7. Use the opposite operation, division, to solve the equation. Divide both sides of the equation by -7:
$$\frac{-7s}{-7} = \frac{49}{-7}$$
$$s = -7$$

3. In the equation $y - 13 = 27$, 13 is subtracted from y. Use the opposite operation, addition, to solve the equation. Add 13 to both sides of the equation:
$$y - 13 + 13 = 27 + 13$$
$$y = 40$$

4. In the equation $\frac{f}{5} = 35$, f is divided by 5. Use the opposite operation, multiplication, to solve the equation. Multiply both sides of the equation by 5:
$$5\left(\frac{f}{5}\right) = 5(35)$$
$$f = 175$$

5. In the equation $54 = 9c$, c is multiplied by 9. Use the opposite operation, division, to solve the equation. Divide both sides of the equation by 9:
$$\frac{54}{9} = \frac{9c}{9}$$
$$c = 6$$

6. In the equation $h - 1 = -1$, 1 is subtracted from h. Use the opposite operation, addition, to solve the equation. Add 1 to both sides of the equation:

$$h - 1 + 1 = -1 + 1$$
$$h = 0$$

7. In the equation $\frac{w}{8} = 6$, w is divided by 8. Use the opposite operation, multiplication, to solve the equation. Multiply both sides of the equation by 8:

$$8\left(\frac{w}{8}\right) = 8(6)$$
$$w = 48$$

8. In the equation $12 = e + 19$, 19 is added to e. Use the opposite operation, subtraction, to solve the equation. Subtract 19 from both sides of the equation:

$$12 - 19 = e + 19 - 19$$
$$e = -7$$

9. In the equation $-2p = -48$, p is multiplied by -2. Use the opposite operation, division, to solve the equation. Divide both sides of the equation by -2:

$$\frac{-2p}{-2} = \frac{-48}{-2}$$
$$p = 24$$

10. In the equation $\frac{3}{4}k = 33$, k is multiplied by $\frac{3}{4}$. Use the opposite operation, division, to solve the equation. Divide both sides of the equation by $\frac{3}{4}$ (which is the same as multiplying by the reciprocal $\frac{4}{3}$):

$$\frac{\frac{3}{4}k}{\frac{3}{4}} = \frac{\frac{33}{3}}{\frac{3}{4}}$$
$$k = 44$$

11. In the equation $12 + a = 2a$, a is on both sides of the equal sign. Dividing both sides by 2 will not get a alone on one side of the equation. Subtracting 12 from both sides will not get a alone, either. Instead, subtract a from both sides of the equation:

$$12 + a - a = 2a - a$$
$$a = 12$$

12. In the equation $9 - v = 0$, if we subtract 9 from both sides, we will have $-v = -9$. We are looking to get v, not $-v$, alone on one side of the equation. Instead, add v to both sides of the equation:

$$9 - v + v = 0 + v$$
$$v = 9$$

Practice 2

1. To find x in terms of y, we must get x alone on one side of the equation, with y on the other side of the equal sign. In the equation $x + 4 = y - 6$, 4 is added to x. Use the opposite operation, subtraction, to get x alone on one side of the equation. Subtract 4 from both sides of the equation:

$$x + 4 - 4 = y - 6 - 4$$
$$x = y - 10$$

The value of x, in terms of y, is $y - 10$.

2. To find x in terms of y, we must get x alone on one side of the equation, with y on the other side of the equal sign. In the equation $2x = 8y$, x is multiplied by 2. Use the opposite operation, division, to get x alone on one side of the equation. Divide both sides of the equation by 2:

$$\frac{2x}{2} = \frac{8y}{2}$$
$$x = 4y$$

The value of x, in terms of y, is $4y$.

3. To find x in terms of y, we must get x alone on one side of the equation, with y on the other side of the equal sign. In the equation $x - 7 = 3y$, 7 is subtracted from x. Use the opposite operation, addition, to get x alone on one side of the equation. Add 7 to both sides of the equation:

$$x - 7 + 7 = 3y + 7$$
$$x = 3y + 7$$

The value of x, in terms of y, is $3y + 7$.

4. To find x in terms of y, we must get x alone on one side of the equation, with y on the other side of the equal sign. In the equation $\frac{x}{9} = y + 1$, x is divided by 9. Use the opposite operation, multiplication, to get x alone on one side of the equation. Multiply both sides of the equation by 9:

$$9\left(\tfrac{x}{9}\right) = 9(y + 1)$$

Use the distributive law to find $9(y + 1)$. Multiply 9 by y and multiply 9 by 1.

$$(9)(y) = 9y, \ (9)(1) = 9$$
$$x = 9y + 9$$

The value of x, in terms of y, is $9y + 9$.

5. To find x in terms of y, we must get x alone on one side of the equation, with y on the other side of the equal sign. In the equation $3x = -12y + 15$, x is multiplied by 3. Use the opposite operation, division, to get x alone on one side of the equation. Divide both sides of the equation by 3:

$$\frac{3x}{3} = \frac{-12y + 15}{3}$$
$$x = -4y + 5$$

The value of x, in terms of y, is $-4y + 5$.

solving multistep
algebraic equations

*Each problem that I solved became a rule which
served afterwards to solve other problems.*
—RENÉ DESCARTES (1596–1650)
FRENCH PHILOSOPHER AND MATHEMATICIAN

In this lesson, you'll learn how to use more than one operation to solve algebraic
equations.

WE OFTEN NEED to use more than one step to solve an equation. We might
have to add and then divide, or add, subtract, and then multiply to find the
value of a variable.

Example
$4x + 2 = 22$

Subtraction alone will not give us the value of x. Division alone will not
give us the value of x, either. We must perform two operations. When you are
solving an equation that requires more than one step, the biggest question is:
Which operation should be done first? The good news is, it does not matter.
Whether we divide first and then subtract, or subtract and then divide, we will
still arrive at the same answer. However, it is usually easier to perform operations in the reverse sequence from the "order of operations." After simplifying,
start with any addition or subtraction, then move on to any multiplication or
division (and then deal with any exponents, if needed).

Let's start by using subtraction. Subtract 2 from both sides of the equation:

$$4x + 2 - 2 = 22 - 2$$
$$4x = 20$$

Now, the equation looks like the equations we solved in the previous lesson. Divide both sides of the equation by 4:

$$\frac{4x}{4} = \frac{20}{4}$$
$$x = 5$$

Solve that equation again, only this time, start with division:

$$\frac{4x + 2}{4} = \frac{22}{4}$$
$$x + \frac{1}{2} = 5\frac{1}{2}$$

$4x + 2$ divided by 4 is $x + \frac{2}{4}$, or $x + \frac{1}{2}$, and 22 divided by 4 is $5\frac{2}{4}$, or $5\frac{1}{2}$. To finish solving the equation, subtract $\frac{1}{2}$ from both sides of the equation:

$$x + \frac{1}{2} - \frac{1}{2} = 5\frac{1}{2} - \frac{1}{2}$$
$$x = 5$$

We found the same answer, $x = 5$, but you might have felt that the problem was a little tougher to solve if you do not like working with fractions. Often, if you divide before adding or subtracting, you will have to work with fractions in order to solve an equation.

. . . .

TIP: If you are solving an equation that requires division, try to add or subtract first and then do the division.

. . . .

Example
$$\frac{g}{6} - 5 = 7$$

In this problem, the variable g is divided by 6, and 5 is subtracted from that quotient. We can either multiply and then add, or add and then multiply. Let's try multiplying first. We must multiply both terms on the left side of the equation:

$$6(\frac{g}{6} - 5) = 6(7)$$
$$g - 30 = 42$$

Now, we can add 30 to both sides of the equation to solve for g:

$$g - 30 + 30 = 42 + 30$$
$$g = 72$$

Would adding before multiplying have made the problem easier to solve? That's up to you. We could have added 5 to both sides of the equation, which would have given us $\frac{g}{6} = 12$. These numbers are a little smaller than 30 and 42, which were the numbers we had after multiplying. Often, if you add or subtract before multiplying, you will have smaller numbers with which to work.

One exception to adding or subtracting first is when an expression on either side of the equal sign can be simplified, such as when a constant or variable is multiplying an expression in parentheses. In that case, it is better to use the distributive law first. Why? You may find that after multiplying, you can combine like terms and make the equation a little easier to solve.

Example
$5(d + 3) - 10 = 6d - 1$

The only way we will be able to either get d alone on the left side of the equation or remove it from the left side of the equation (and get it alone on the right side) is to multiply $(d + 3)$ by 5. Even if we add 1 to both sides first or add 10 to both sides first, we will have to multiply next. Start by multiplying $5(d + 3)$:

$$5(d + 3) = 5d + 15$$

The equation is now:

$$5d + 15 - 10 = 6d - 1$$

Combine the constants on the left side by subtracting 10 from 15:

$$5d + 5 = 6d - 1$$

We have a few choices for our next step. We could add 1 to both sides of the equation, we could subtract 5 from both sides, or we could subtract $5d$ from both sides. It does not matter which we do next, so let's add 1 to both sides (since adding 1 is easy!).

$$5d + 5 + 1 = 6d - 1 + 1$$
$$5d + 6 = 6d$$

Now, subtract $5d$ from both sides of the equation:

$$5d - 5d + 6 = 6d - 5d$$
$$6 = d$$

We can check our work by substituting our answer back into the original equation. Substitute 6 for d in the last example:

$$5(d + 3) - 10 = 6d - 1$$
$$5(6 + 3) - 10 = 6(6) - 1$$
$$5(9) - 10 = 6(6) - 1$$
$$45 - 10 = 36 - 1$$
$$35 = 35$$

It is true that 35 equals 35, so we know that our answer is correct. That example took many steps: We multiplied, subtracted, added, and then subtracted again. Some equations might take even more steps to solve. Every step we take should bring us a little closer to the value of the variable. Most steps should remove at least one term from one side of the equation. By reducing the total number of terms, we move closer to getting the equation in the form $x =$ (some number).

Practice

Find the value of the variable in each equation.

1. $5u + 3 = 48$

2. $3f = 4 + f$

3. $\frac{a}{7} - 2 = 6$

4. $-2p + 3 = 13$

5. $\frac{c}{2} + 4c = -18$

6. $6r + 4 = -r - 24$

7. $4(q - 3) = 16$

8. $10(\frac{v}{2} + 1) = 6v + 4$

9. $\frac{e}{4} + 10 = e - 8$

10. $12b + 21 = -2b - 21$

11. $\frac{-2z}{3} + 8 = 2$

12. $5x - 20 = \frac{x}{3} + 8$

13. $-3(t - 11) = 4t - 2$

14. $\frac{1}{5}(x + 10) = 26 - x$

15. $9(4p + 12) = 15p + 3$

ANSWERS

Practice

1. The equation $5u + 3 = 48$ shows multiplication and addition. We will need to use their opposites, division and subtraction, to find the value of u. Subtract first:

$$5u + 3 - 3 = 48 - 3$$
$$5u = 45$$

Because u is multiplied by 5, divide both sides of the equation by 5:

$$\frac{5u}{5} = \frac{45}{5}$$
$$u = 9$$

2. The equation $3f = 4 + f$ shows multiplication and addition. We will need to use their opposites, division and subtraction, to find the value of f. Subtracting 4 from both sides of the equation will not reduce the number of terms, but subtracting f from both will:

$$3f - f = 4 + f - f$$
$$2f = 4$$

Because f is multiplied by 2, divide both sides of the equation by 2:

$$\frac{2f}{2} = \frac{4}{2}$$
$$f = 2$$

3. The equation $\frac{a}{7} - 2 = 6$ shows division and subtraction. We will need to use their opposites, multiplication and addition, to find the value of a. If we add first, we will have smaller numbers with which to work:

$$\frac{a}{7} - 2 + 2 = 6 + 2$$
$$\frac{a}{7} = 8$$

Because a is divided by 7, multiply both sides of the equation by 7:

$$7(\tfrac{a}{7}) = 7(8)$$
$$a = 56$$

4. The equation $-2p + 3 = 13$ shows multiplication and addition. We will need to use their opposites, division and subtraction, to find the value of p. Subtract first:

$$-2p + 3 - 3 = 13 - 3$$
$$-2p = 10$$

Because p is multiplied by -2, divide both sides of the equation by -2:

$$\frac{-2p}{-2} = \frac{10}{-2}$$
$$p = -5$$

5. The equation $\frac{c}{2} + 4c = -18$ shows division, addition, and multiplication. Because we have like terms on the left side of the equation, we can simplify by adding:

$$\frac{c}{2} + 4c = 4\tfrac{1}{2}c$$
$$4\tfrac{1}{2}c = -18$$

Because c is multiplied by $4\tfrac{1}{2}$, divide both sides of the equation by $4\tfrac{1}{2}$:

$$\frac{4.5c}{4.5} = \frac{-18}{4.5}$$
$$c = -4$$

6. The equation $6r + 4 = -r - 24$ shows multiplication, addition, and subtraction. Both sides of the equation contain an r term and both sides contain constants. We must get the variable alone on one side and a constant alone on the other. We can remove r from the right side by adding r to both sides of the equation:

$$6r + r + 4 = -r + r - 24$$
$$7r + 4 = -24$$

Now the equation shows addition and multiplication, so we must use subtraction and division to solve it. Subtract 4 from both sides of the equation:

$$7r + 4 - 4 = -24 - 4$$
$$7r = -28$$

Because r is multiplied by 7, divide both sides of the equation by 7:

$$\frac{7r}{7} = \frac{-28}{7}$$

$$r = -4$$

7. The equation $4(q - 3) = 16$ has a constant multiplying an expression. The first step to solving this equation is to simplify the left side of the equation using the distributive law. Multiply q and -3 by 4:

$$4(q - 3) = 4q - 12$$

The equation is now:

$$4q - 12 = 16$$

The equation shows subtraction and multiplication, so we must use addition and division to solve it. Add 12 to both sides of the equation:

$$4q - 12 + 12 = 16 + 12$$

$$4q = 28$$

Because q is multiplied by 4, divide both sides of the equation by 4:

$$\frac{4q}{4} = \frac{28}{4}$$

$$q = 7$$

8. The equation $10(\frac{v}{2} + 1) = 6v + 4$ has a constant multiplying an expression. The first step to solving this equation is to simplify using the distributive law. Multiply $\frac{v}{2}$ and 1 by 10:

$$10(\frac{v}{2} + 1) = 5v + 10$$

The equation is now:

$$5v + 10 = 6v + 4$$

We must get the variable alone on one side and a constant alone on the other. We can remove v from the left side by subtracting $5v$ from both sides of the equation:

$$5v - 5v + 10 = 6v - 5v + 4$$

$$10 = v + 4$$

To get v alone on the right side of the equation, subtract 4 from both sides:

$$10 - 4 = v + 4 - 4$$

$$6 = v$$

9. The equation $\frac{e}{4} + 10 = e - 8$ shows division, addition, and subtraction. The variable e appears on both sides of the equation, and a constant is on both sides of the equation. Start by subtracting $\frac{e}{4}$ from both sides. This will leave us with just one e term:

$$\frac{e}{4} - \frac{e}{4} + 10 = e - \frac{e}{4} - 8$$

$$10 = \frac{3}{4}e - 8$$

Because 8 is subtracted from $\frac{3}{4}e$, add 8 to both sides of the equation:

$$10 + 8 = \frac{3}{4}e - 8 + 8$$

$$18 = \frac{3}{4}e$$

Finally, because e is multiplied by $\frac{3}{4}$, divide both sides of the equation by $\frac{3}{4}$:

$$\frac{18}{\frac{3}{4}} = \frac{\frac{3}{4}e}{\frac{3}{4}}$$

$$24 = e$$

Instead of dividing by $\frac{3}{4}$, we could have also multiplied by its reciprocal, $\frac{4}{3}$. Both operations get e alone on one side of the equation and give us an answer of 24.

10. The equation $12b + 21 = -2b - 21$ shows multiplication, addition, and subtraction. Both sides of the equation contain a b term and both sides contain constants. We must get the variable alone on one side and a constant alone on the other. We can remove b from the right side by adding $2b$ to both sides of the equation:

$$12b + 2b + 21 = -2b + 2b - 21$$
$$14b + 21 = -21$$

Now, the equation shows addition and multiplication, so we must use subtraction and division to solve it. Subtract 21 from both sides of the equation:

$$14b + 21 - 21 = -21 - 21$$
$$14b = -42$$

Because b is multiplied by 14, divide both sides of the equation by 14:

$$\frac{14b}{14} = \frac{-42}{14}$$
$$b = -3$$

11. The equation $\frac{-2z}{3} + 8 = 2$ shows multiplication and addition. We will need to use their opposites, division and subtraction, to find the value of z. Subtract 8 from both sides of the equation:

$$\frac{-2z}{3} + 8 - 8 = 2 - 8$$
$$\frac{-2z}{3} = -6$$

We can divide both sides of the equation by $\frac{-2}{3}$, which is the same as multiplying by its reciprocal $\frac{-2}{3}$. Let's multiply:

$$\left(\frac{-3}{2}\right) - \frac{-2z}{3} = -6\left(\frac{-3}{2}\right)$$
$$z = 9$$

12. The equation $5x - 20 = \frac{x}{3} + 8$ shows multiplication, subtraction, division, and addition. Both sides of the equation contain an x term and both sides contain constants. We must get the variable alone on one side and a constant alone on the other. We can remove x from the right side by subtracting $\frac{x}{3}$ from both sides of the equation:

$$5x - \frac{x}{3} - 20 = \frac{x}{3} - \frac{x}{3} + 8$$
$$\frac{14}{3x} - 20 = 8$$

Now, the equation shows multiplication and subtraction, so we must use addition and division to solve it. Add 20 to both sides of the equation:

$$\frac{14}{3x} - 20 + 20 = 8 + 20$$

$$\frac{14}{3x} = 28$$

Divide both sides of the equation by $\frac{14}{3}$, or multiply by its reciprocal $\frac{3}{14}$, to get x alone on the left side:

$$\frac{3}{14}\left(\frac{14}{3x}\right) = \frac{3}{14}(28)$$

$$x = 6$$

13. The equation $-3(t - 11) = 4t - 2$ has a constant multiplying an expression. The first step to solving this equation is to use the distributive law. Multiply t and -11 by -3:

$$-3(t - 11) = -3t + 33$$

The equation is now:

$$-3t + 33 = 4t - 2$$

We must get the variable alone on one side and a constant alone on the other. We can remove t from the left side by adding $3t$ to both sides of the equation:

$$-3t + 3t + 33 = 4t + 3t - 2$$

$$33 = 7t - 2$$

Next, remove -2 from the right side of the equation by adding 2 to both sides:

$$33 + 2 = 7t - 2 + 2$$

$$35 = 7t$$

Because t is multiplied by 7, divide both sides of the equation by 7:

$$\frac{35}{7} = \frac{7t}{7}$$

$$5 = t$$

14. The equation $\frac{1}{5}(x + 10) = 26 - x$ has a constant multiplying an expression. The first step to solving this equation is to use the distributive law. Multiply x and 10 by $\frac{1}{5}$:

$$\frac{1}{5}(x + 10) = \frac{1}{5}x + 2$$

The equation is now:

$$\frac{1}{5}x + 2 = 26 - x$$

We must get the variable alone on one side and a constant alone on the other. We can remove x from the right side by adding x to both sides of the equation:

$$\frac{1}{5}x + x + 2 = 26 - x + x$$

$$\frac{6}{5}x + 2 = 26$$

Then, subtract 2 from both sides:

$$\frac{6}{5}x + 2 - 2 = 26 - 2$$

$$\frac{6}{5}x = 24$$

Multiply both sides of the equation by $\frac{5}{6}$ to get x alone on the left side:

$$\frac{5}{6}(\frac{6}{5}x) = \frac{5}{6}(24)$$

$$x = 20$$

15. The equation $9(4p + 12) = 15p + 3$ has a constant multiplying an expression. The first step to solving this equation is to use the distributive law. Multiply $4p$ and 12 by 9:

$$9(4p + 12) = 36p + 108$$

The equation is now:

$$36p + 108 = 15p + 3$$

We must get the variable alone on one side and a constant alone on the other. We can remove p from the right side by subtracting $15p$ from both sides of the equation:

$$36p - 15p + 108 = 15p - 15p + 3$$

$$21p + 108 = 3$$

Next, remove 108 from the left side of the equation by subtracting 108 from both sides:

$$21p + 108 - 108 = 3 - 108$$

$$21p = -105$$

Because p is multiplied by 21, divide both sides of the equation by 21:

$$\frac{21p}{21} = \frac{-105}{21}$$

$$p = -5$$

radicals

Math is radical!
—BUMPER STICKER

In this lesson, you'll learn how to work with radicals and fractional exponents and how to solve equations with radicals.

WE FIRST LOOKED at radicals in Lesson 8. A **radical** is a root of a quantity. You can think of radicals and exponents as opposites. If we raise a quantity, x, to the fourth power, we multiply it four times: $x^4 = (x)(x)(x)(x)$. If we take the fourth root of x, we are looking for the number that, when multiplied four times, is equal to x. We show the fourth root of x as $\sqrt[4]{x}$. The $\sqrt{}$ symbol is called the radical symbol. The quantity under the radical symbol is called the **radicand**. The number just outside the radical symbol is the root that must be taken. In $\sqrt[4]{x}$, x is the radicand and 4 is the root. If no number appears outside the radical symbol, then 2 is the root.

Let's look at an example with numbers: $2^3 = (2)(2)(2) = 8$. The third root of 8 is 2, because 2 is the number that, when multiplied three times, is equal to 8: $\sqrt[3]{8} = 2$.

A radicand can have an exponent. The square root of x^4, $\sqrt{x^4}$, is x^2, because $(x^2)(x^2) = x^4$.

TIP: If the root of a radical is the same as the exponent of the base of the radicand, and the root is positive, the expression can be simplified to the base of the radicand. For example, the fifth root of x to the fifth power is x. $\sqrt[5]{x^5} = x$. For positive values of x, the tenth root of x to the tenth power is x, and so on.

We cannot find even roots of negative numbers. For example, we cannot find the square root of a negative number, because no number multiplied by itself is negative. A positive number multiplied by a positive number gives you a positive product, and a negative number multiplied by a negative number gives you a positive product. We can find the odd roots of some negative numbers. For instance, $\sqrt[3]{-x^3} = -x$, because $(-x)(-x)(-x) = -x^3$.

If a coefficient appears in the radicand, we must take the root of the coefficient as well as the base. $\sqrt{16z^8} = 4z^4$, because $(4z^4)(4z^4) = 16z^8$. The square root of an even exponent is always half the exponent.

A radical itself can have a coefficient. It is written to the left of the radical symbol. $5\sqrt{b^6}$ is 5 times the square root of b^6. Half of 6 is 3, so the square root of b^6 is b^3. Five times b^3 is $5b^3$.

The exponent of a variable can be written as a fraction, as we saw in Lesson 8. The numerator of a fractional exponent is the power to which the variable is raised. The denominator of the fractional exponent is the root to take of the variable. $g^{\frac{7}{3}}$ is equal to $\sqrt[3]{g^7}$, which is the third root of g to the seventh power. g^7 does not have a third root that we can easily calculate, so we cannot simplify this expression any further. We were, though, able to find the square root of b^6. The reason why becomes clearer when we write the square root of b to the sixth power as a fractional exponent. Because we are raising b to the sixth power, the numerator of the fraction is 6, and since we are taking the square, or second, root of b, the denominator of the fraction is 2: $b^{\frac{6}{2}}$. The fraction $\frac{6}{2}$ reduces to 3, just as the square root of b to the sixth power is equal to b^3.

TIP: Try to memorize the squares of numbers up to 20, and their square roots. This will help you work with exponents and radicals faster.

Practice 1

Simplify each expression as much as possible.

1. $\sqrt{m^{10}}$

2. $c^{\frac{4}{3}}$

3. $\sqrt[3]{j^{15}}$

4. $2\sqrt{25v^2}$

5. $-4\sqrt[3]{27k^6}$

SOLVING EQUATIONS WITH RADICALS

Now that we know how to work with radicals, we can use them to help us solve equations. So far, we have used addition, subtraction, multiplication, and division to solve equations. Sometimes, we will need to raise both sides of an equation to a power, or take a root of both sides of an equation in order to find our answer.

> **Example**
> $x^2 = 64$

How can we find the value of x? Addition, subtraction, multiplication, and division cannot help us. However, we can get x alone on the left side of the equation if we take the square root of both sides of the equation. Why the square root? Because the exponent of x is 2. If the exponent of x was 5, we would take the fifth root of x to get x alone. Remember the tip we learned earlier: If the exponent and root of a base are the same, the term can be simplified to just the base or its absolute value.

$$\sqrt{x^2} = \sqrt{64}$$

The square root of x^2 is the absolute value of x and the square root of 64 is 8, since $(8)(8) = 64$. We are left with $x = 8$. This isn't our only answer, though. It's true that the square root of 64 is 8, but there is another number that, when squared, equals 64: –8. Remember, $(-8)(-8) = 64$.

Our answers are $x = 8, -8$.

Example

$4x^2 = 144$

To solve for x, we start by dividing both sides of the equation by 4.

$$\frac{4x^2}{4} = \frac{144}{4}$$
$$x^2 = 36$$

Now, we can take the plus and minus square roots of 36. Because $(6)(6) = 36$ and $(-6)(-6) = -36$, our answers are $x = 6, -6$.

We can also use exponents to help us simplify radicals. To remove the radical symbol from an equation, raise it to an exponent that is equal to the root of the radical.

Example

$\sqrt[4]{r} = 3$

Because the fourth root of r is equal to 3, raise both sides of the equation to the fourth power to remove the radical symbol.

$$(\sqrt[4]{r})^4 = 3^4$$
$$r = 81$$

Practice 2

Solve the following equations.

1. $h^3 = 64$

2. $c^2 - 8 = 1$

3. $5n^2 = 80$

4. $\sqrt{3p} = 9$

5. $\sqrt[3]{5d} = 5$

ANSWERS

Practice 1

1. The square root of m^{10} is the quantity that, when multiplied by itself, is equal to m^{10}. To find the square root of a base with an even exponent, divide the exponent by 2: $10 \div 2 = 5$. $(m^5)(m^5) = m^{10}$, which is why $\sqrt{m^{10}} = m^5$.

2. The numerator of a fractional exponent is the power to which the variable is raised. The denominator of the fractional exponent is the root to take of the variable. $c^{\frac{4}{3}}$ is equal to $\sqrt[3]{c^4}$, which is the third root of c to the fourth power. There is no whole exponent of c, such that we could multiply it by itself three times and arrive at c^4, so this expression cannot be simplified any further.

3. The third root of j^{15} is the quantity that, when multiplied three times, is equal to j^{15}. To find the third root of a base, divide the exponent of the base by 3: $15 \div 3 = 5$. $(j^5)(j^5)(j^5) = j^{15}$, which is why $\sqrt[3]{j^{15}} = j^5$.

4. The square root of $25v^2$ is the quantity that, when multiplied by itself, is equal to $25v^2$. Find the square root of the coefficient and the square root of the base with its exponent. The square root of 25 is 5, because $(5)(5) = 25$. To find the square root of a base with an even exponent, divide the exponent by 2: $2 \div 2 = 1$. $(v)(v) = v^2$, which is why $\sqrt{v^2} = v$. Therefore, $\sqrt{25v^2} = 5v$. Multiply the coefficient of the radical, 2, by $5v$: $2(5v) = 10v$.

5. The third root of $27k^6$ is the quantity that, when multiplied three times, is equal to $27k^6$. Find the third root of the coefficient and the third root of the base with its exponent. The third root of 27 is 3, because $(3)(3)(3) = 27$. To find the third root of a base, divide the exponent of the base by 3: $6 \div 3 = 2$. $(k^2)(k^2)(k^2) = k^6$, which is why $\sqrt[3]{k^6} = k^2$. Therefore, $\sqrt[3]{27k^6} = 3k^2$. Multiply the coefficient of the radical, -4, by $3k^2$: $-4(3k^2) = -12k^2$.

Practice 2

1. In the equation $h^3 = 64$, h is raised to the third power. To get h alone on the left side of the equation, we must take the third root of both sides of the equation. The third root of h^3 is h, because $(h)(h)(h) = h^3$. The third root of 64 is 4, because $(4)(4)(4) = 64$.

$$\sqrt[3]{h^3} = \sqrt[3]{64}$$
$$h = 4$$

2. In the equation $c^2 - 8 = 1$, c is raised to the second power and 8 is subtracted from that square. First, add 8 to both sides of the equation so that the variable is on one side of the equation and the constant is on the other side:

$$c^2 - 8 + 8 = 1 + 8$$
$$c^2 = 9$$

To get c alone on the left side of the equation, we must take the square root of both sides of the equation. The square root of c^2 is c, because $(c)(c) = c^2$. Because we are taking the square root, an even root, of a constant on the right side of the equation, we must take the plus and minus square roots. The positive square root of 9 is 3, since $(3)(3) = 9$. The negative square root of 9 is -3, because $(-3)(-3) = 9$.

$$\sqrt{c^2} = \pm\sqrt{9}$$
$$c = 3, -3$$

3. In the equation $5n^2 = 80$, n is raised to the second power and then multiplied by 5. First, divide both sides of the equation by 5 so that the variable is on one side of the equation and the constant is on the other side:

$$\frac{5n^2}{5} = \frac{80}{5}$$
$$n^2 = 16$$

To get n alone on the left side of the equation, we must take the square root of both sides of the equation. The square root of n^2 is n, because $(n)(n) = n^2$. Because we are taking the square root, an even root, of a constant on the right side of the equation, we must take the plus and minus square roots. The positive square root of 16 is 4, because $(4)(4) = 16$. The negative of the square root of 16 is -4, because $(-4)(-4) = 16$.

$$\sqrt{n^2} = \pm\sqrt{16}$$
$$n = 4, -4$$

4. In the equation $\sqrt{3p} = 9$, the square root of $3p$ is equal to 9. To remove the radical symbol from the left side of the equation, we must raise both sides of the equation to the second power.

$$(\sqrt{3p})^2 = (9)^2$$
$$3p = 81$$

Now the equation looks like the ones we have solved in Lessons 9 and 10. Because p is multiplied by 3, divide both sides of the equation by 3:

$$\frac{3p}{3} = \frac{81}{3}$$
$$p = 27$$

5. In the equation $\sqrt[3]{5d} = 5$, the third root of $5d$ is equal to 5. To remove the radical symbol from the left side of the equation, we must raise both sides of the equation to the third power.

$$(\sqrt[3]{5d})^3 = (5)^3$$

$$5d = 125$$

Because d is multiplied by 5, divide both sides of the equation by 5:

$$\frac{5d}{5} = \frac{125}{5}$$

$$d = 25$$

slope-intercept form

In this lesson, you'll learn about the slope and y-intercept of the equation of a line, how to put an equation in slope-intercept form, and how to find the slope and y-intercept of a line.

THE VARIABLES x and y can be used to form the equation of a line. A horizontal line takes the form $y = c$, where c is any constant, such as 0, 3, or –4. A vertical line takes the form $x = c$, such as $x = 2$ or $x = -14$. The equations of all other lines contain both x and y. These lines have a slope and a y-intercept. **Slope** is the change in the y values between two points on a line divided by the change in the x values of those points. The **y-intercept** of a line is the y value of the point where the line crosses the y-axis.

We write the equation of a line in slope-intercept form so that it is easy to spot the slope and y-intercept just by looking at the equation. **Slope-intercept form** is $y = mx + b$, where m is the slope of the line and b is the y-intercept.

Example
$y = 2x + 5$

This equation is already in slope-intercept form. We can tell that it is in slope-intercept form because y is alone on one side of the equal sign, and the other side of the equation contains no more than two terms, with x appearing in no more than one term. The coefficient of x is 2, which means that the slope of the line is 2. The constant that is added to $2x$ is 5, which means that 5 is the y-intercept of the line.

If no constant is added to the x term, then the y-intercept of the line is 0. The equations $y = 4x$, $y = -3x$, and $y = \frac{1}{2}x$ all have a y-intercept of 0.

It is also possible for a line to have a slope of 0. If x does not appear in the equation of a line, then the line has a slope of 0. The lines $y = 4$, $y = -10$, and $y = 0$ are all in slope-intercept form, and they all have slopes of 0.

....

TIP: Vertical lines—lines in the form $x = c$—do not have a slope of 0. They have no slope at all. These equations are not in slope-intercept form, and they cannot be put in slope-intercept form because they do not contain the variable y.

....

Practice 1

For each equation, find the slope and the y-intercept.

1. $y = 9x + 4$

2. $y = \frac{2}{3}x - 6$

3. $y = -5x$

4. $y = 12$

5. $y = x$

PUTTING AN EQUATION IN SLOPE-INTERCEPT FORM

If the equation of a line is not in slope-intercept form, we must perform one or more operations to get y alone on one side of the equation, with x in no more than one term on the other side of the equation.

The equation $y + 9 = 3x$ is not in slope-intercept form, because the constant, 9, is on the same side of the equation as y. To remove 9 from the left side of the equation, subtract 9 from both sides of the equation: $y + 9 - 9 = 3x - 9$, and $y = 3x - 9$. The equation is now in slope-intercept form, and we can see that the slope is 3 and the y-intercept is –9.

Example

$3y = 6x + 30$

In this equation, y is in the only term on the left side of the equation, but it is multiplied by 3. To get y alone, divide both sides of the equation by 3. We must divide both terms on the right side of the equation by 3:

$$\frac{3y}{3} = \frac{6x + 30}{3}$$
$$y = 2x + 10$$

We can see now that the slope of the line is 2 and the y-intercept is 10.

Example

$2x + 4y - 8 = 12$

It will take a few steps to put this equation in slope-intercept form. We must move the x term and the constant to the right side of the equation, so that y can be alone on the left side. Subtract $2x$ from both sides of the equation and add 8 to both sides:

$$2x - 2x + 4y - 8 + 8 = 12 - 2x + 8$$
$$4y = -2x + 20$$

Because y is multiplied by 4, divide both sides of the equation by 4 to get y alone:

$$\frac{4y}{4} = \frac{-2x + 20}{4}$$
$$y = \frac{-1}{2}x + 5$$

The slope of this line is $-\frac{1}{2}$ and the y-intercept is 5.

Practice 2

For each equation, find the slope and the y-intercept.

1. $y - 3 = x + 4$

2. $5y = -15x - 25$

3. $y = 7(x + 1)$

4. $x + y = 0$

5. $x = 4y - 24$

PARALLEL AND PERPENDICULAR LINES

We can tell if two lines are parallel to each other or perpendicular to each other just by looking at the slopes of the lines. **Parallel lines** are lines that have the same slope. The lines each extend forever and never cross. **Perpendicular lines** are lines that cross at right angles. The slopes of perpendicular lines are negative reciprocals of each other. What does this mean? The reciprocal of an integer is 1 divided by that integer. The reciprocal of 5 is $\frac{1}{5}$. To find the reciprocal of a fraction, just switch the numerator and denominator. The reciprocal of $\frac{2}{7}$ is $\frac{7}{2}$. To find the negative reciprocal of an integer, divide 1 by the integer and change the sign of the integer. The negative reciprocal of 2 is $-\frac{1}{2}$, because 1 divided by 2 is $\frac{1}{2}$, and because the sign of 2 was positive, we change it to negative when we find its negative reciprocal.

The line $y = 3x + 5$ is parallel to the line $y = 3x - 2$, because they have the same slope, 3. The only difference between the two equations is their y-intercepts. If the lines had the same y-intercepts, they would be the exact same line. Any line with a slope of 3 is parallel to these lines.

The line $y = 3x + 5$ is perpendicular to the line $y = -\frac{1}{3}x + 1$, because 3 and $-\frac{1}{3}$ are negative reciprocals of each other. Any line with a slope of $-\frac{1}{3}$ is perpendicular to the line $y = 3x + 5$.

Let's look at one more example. What lines are parallel to the line $y = \frac{3}{8}x + 1$ and what lines are perpendicular to it? Any line with a slope of $\frac{3}{8}$, such as $y = \frac{3}{8}x + 20$, is parallel to $y = \frac{3}{8}x + 1$. The negative reciprocal of $\frac{3}{8}$ is $-\frac{8}{3}$. Any line with a slope of $-\frac{8}{3}$, such as $y = -\frac{8}{3}x$, is perpendicular to $y = \frac{3}{8}x + 1$.

> **TIP:** When finding parallel lines or perpendicular lines, ignore the y-intercept. Only the slope determines whether two lines are parallel, perpendicular, or neither.

Practice 3

Write the equation of a line that is parallel to each of the following equations, and write the equation of a line that is perpendicular to each equation.

1. $y = 2x - 9$

2. $y = -7x$

3. $y = \frac{7}{9}x - \frac{3}{8}$

4. $y = -\frac{1}{4}x + 34$

5. $y = x$

ANSWERS

Practice 1

1. The equation $y = 9x + 4$ is in slope-intercept form, because y is alone on one side of the equation, and the other side of the equation contains no more than two terms, with x appearing in no more than one of those terms. The slope is the coefficient of x, so the slope of this line is 9. The y-intercept is the constant in the equation, which is 4.
2. The equation $y = \frac{2}{3}x - 6$ is in slope-intercept form. The slope of the line is the coefficient of x, $\frac{2}{3}$, and the y-intercept of the line is the constant in the equation, -6.
3. The equation $y = -5x$ is in slope-intercept form. The slope of the line is the coefficient of x, -5. There is no constant in this equation, so the y-intercept of the line is 0.

4. The equation $y = 12$ is in slope-intercept form. There is no x term in the equation, so the slope of the line is 0. The y-intercept of the line is the constant in the equation, 12.

5. The equation $y = x$ is in slope-intercept form. The slope of the line is the coefficient of x, 1. There is no constant in this equation, so the y-intercept of the line is 0.

Practice 2

1. The equation $y - 3 = x + 4$ is not in slope-intercept form, because y is not alone on one side of the equation. Because 3 is subtracted from y, we must use the opposite operation, addition, to remove –3 from the left side of the equation. Add 3 to both sides of the equation:
$$y - 3 + 3 = x + 4 + 3$$
$$y = x + 7$$
The equation is now in slope-intercept form. The slope is the coefficient of x, so the slope of this line is 1. The y-intercept is the constant in the equation, which is 7.

2. The equation $5y = -15x - 25$ is not in slope-intercept form, because y is not alone on one side of the equation. Because y is multiplied by 5, we must divide both sides of the equation by 5:
$$\frac{5y}{5} = \frac{-15x - 25}{5}$$
$$y = -3x - 5$$
The equation is now in slope-intercept form. The slope is the coefficient of x, so the slope of this line is –3. The y-intercept is the constant in the equation, which is –5.

3. The equation $y = 7(x + 1)$ is not in slope-intercept form. Although y is alone on one side of the equation, there are parentheses around two terms on the right side. Use the distributive law to simplify the right side of the equation. Multiply x and 1 by 7:
$$7(x + 1) = 7x + 7$$
The equation is now in slope-intercept form:
$$y = 7x + 7$$
The slope is the coefficient of x, so the slope of this line is 7. The y-intercept is the constant in the equation, which is 7.

4. The equation $x + y = 0$ is not in slope-intercept form, because y is not alone on one side of the equation. Because x is added to y, we must use the

opposite operation, subtraction, to remove x from the left side of the equation. Subtract x from both sides of the equation:

$$x - x + y = 0 - x$$
$$y = -x$$

The equation is now in slope-intercept form. The slope is the coefficient of x, so the slope of this line is -1. There is no constant in this equation, so the y-intercept of the line is 0.

5. The equation $x = 4y - 24$ is not in slope-intercept form, because y is not alone on one side of the equation. In the equation, y is multiplied by 4, and then 24 is subtracted from that term. First, add 24 to both sides of the equation:

$$x + 24 = 4y - 24 + 24$$
$$x + 24 = 4y$$

Because y is multiplied by 4, divide both sides of the equation by 4:

$$\frac{x + 24}{4} = \frac{4y}{4}$$
$$y = \tfrac{1}{4}x + 6$$

The equation is now in slope-intercept form. The slope is the coefficient of x, so the slope of this line is $\tfrac{1}{4}$. The y-intercept is the constant in the equation, which is 6.

Practice 3

1. The slope of the line $y = 2x - 9$ is 2, because the line is in slope-intercept form, and the coefficient of x is 2. Any line with a slope of 2, such as $y = 2x + 1$, is parallel to the line $y = 2x - 9$.

    Because the slope of the line $y = 2x - 9$ is 2, lines that are perpendicular to this line will have slopes that are the negative reciprocal of 2. To find the negative reciprocal of 2, divide 1 by 2, and change the sign from positive to negative. The negative reciprocal of 2 is $-\tfrac{1}{2}$. Any line with a slope of $-\tfrac{1}{2}$, such as $y = -\tfrac{1}{2}x + 1$, is perpendicular to the line $y = 2x - 9$.

2. The slope of the line $y = -7x$ is -7, because the line is in slope-intercept form, and the coefficient of x is -7. Any line with a slope of -7, such as $y = -7x + 1$, is parallel to the line $y = -7x$.

    Because the slope of the line $y = -7x$ is -7, lines that are perpendicular to this line will have slopes that are the negative reciprocal of -7. To find the negative reciprocal of -7, divide 1 by -7, and change the sign from negative to positive. The negative reciprocal of -7 is $\tfrac{1}{7}$. Any line with a slope of $\tfrac{1}{7}$, such as $y = \tfrac{1}{7}x + 1$, is perpendicular to the line $y = -7x$.

3. The slope of the line $y = \frac{7}{9}x - \frac{3}{8}$ is $\frac{7}{9}$, because the line is in slope-intercept form, and the coefficient of x is $\frac{7}{9}$. Any line with a slope of $\frac{7}{9}$, such as $y = \frac{7}{9}x + 1$, is parallel to the line $y = \frac{7}{9}x - \frac{3}{8}$.

Because the slope of the line $y = \frac{7}{9}x - \frac{3}{8}$ is $\frac{7}{9}$, lines that are perpendicular to this line will have slopes that are the negative reciprocal of $\frac{7}{9}$. To find the negative reciprocal of $\frac{7}{9}$, switch the numerator and the denominator of the fraction, and change the sign from positive to negative. The negative reciprocal of $\frac{7}{9}$ is $-\frac{9}{7}$. Any line with a slope of $-\frac{9}{7}$, such as $y = -\frac{9}{7}x + 1$, is perpendicular to the line $y = \frac{7}{9}x - \frac{3}{8}$.

4. The slope of the line $-\frac{1}{4}x + 34$ is $-\frac{1}{4}$, because the line is in slope-intercept form, and the coefficient of x is $-\frac{1}{4}$. Any line with a slope of $-\frac{1}{4}$, such as $y = -\frac{1}{4}x + 1$, is parallel to the line $-\frac{1}{4}x + 34$.

Because the slope of the line $-\frac{1}{4}x + 34$ is $-\frac{1}{4}$, lines that are perpendicular to this line will have slopes that are the negative reciprocal of $-\frac{1}{4}$. To find the negative reciprocal of $-\frac{1}{4}$, switch the numerator and the denominator of the fraction, and change the sign from negative to positive. The negative reciprocal of $-\frac{1}{4}$ is $\frac{4}{1}$, or 4. Any line with a slope of 4, such as $y = 4x + 1$, is perpendicular to the line $-\frac{1}{4}x + 34$.

5. The slope of the line $y = x$ is 1, because the line is in slope-intercept form, and the coefficient of x is 1. Any line with a slope of 1, such as $y = x + 1$, is parallel to the line $y = x$.

Because the slope of the line $y = x$ is 1, lines that are perpendicular to this line will have slopes that are the negative reciprocal of 1. To find the negative reciprocal of 1, divide 1 by 1, and change the sign from positive to negative. Because 1 divided by 1 is 1, the negative reciprocal of 1 is -1. Any line with a slope of -1, such as $y = -x + 1$, is perpendicular to the line $y = x$.

input/output tables

Mathematical discoveries, small or great
are never born of spontaneous generation.
—JULES HENRI POINCARÉ (1854–1912)
FRENCH MATHEMATICIAN

In this lesson, you'll learn about input/output tables: how to find the rule of a table and how to find a missing value in a table.

THE VARIABLES IN THE EQUATION of a line have a relationship between each other. As the value of one variable changes, so does the value of the other variable. In the equation $y = 4x - 1$, when $x = 3$, y is equal to $4(3) - 1 = 12 - 1 = 11$. If x is equal to 4, then y is equal to $4(4) - 1 = 16 - 1 = 15$.

For any equation of a line, we can make an **input/output table** to show the relationship between x and y. The input/output table of $y = 4x - 1$ shows the value of y for different values of x:

x	$y = 4x - 1$	y
0	$4(0) - 1$	−1
1	$4(1) - 1$	3
2	$4(2) - 1$	7
3	$4(3) - 1$	11
4	$4(4) - 1$	15

The first column shows the value of x. We say that x is the independent variable, because it determines the value of y, and we say that y is the dependent variable, because we find its value based on the value of x. The middle column of the table shows the substitution of a number for x, and the last column shows the value of y after those calculations have been performed. Most input/output tables do not show the middle column, just the values of x and y.

The equation used for this table is listed right at the top of the table, but it usually is not given to us; we will need to figure out what the equation is based on the values in the table.

x	y
0	5
1	6
2	7
3	8
4	9

This input/output table is the kind of table we will usually see—just 4 or 5 x values with their corresponding y values. In Lesson 12, we learned that the slope of a line is the change in the y-values between two points on a line divided by the change in the x-values of those points. Because we were given equations in that lesson, we could find the slope of a line by putting the equation in slope-intercept form. Now that we have an input/output table, we can figure out the slope of a line ourselves!

We can find the slope of a line using any two points on the line. We will use the first two rows of this input/output table. First, find the difference between the y values: $6 - 5 = 1$. Next, find the difference between the x values: $1 - 0 = 1$. Now, divide the difference between the y values by the difference between the x values: $1 \div 1 = 1$. The slope of this equation is 1.

We have the slope of the equation, and now we need the y-intercept. Remember: Slope-intercept form is $y = mx + b$, where m is the slope of the line and b is the y-intercept. We have plenty of x and y values, and we know the slope of the line is 1. We can substitute these values into the equation and solve for b. Any row from the table will do. Let's use the last row, where x is 4 and y is 9. Substitute these values into the equation $y = mx + b$:

$$9 = 1(4) + b$$
$$9 = 4 + b$$

Subtract 4 from both sides of the equation:

$$9 - 4 = 4 + b - 4$$
$$5 = b$$

The y-intercept of the line is 5. Now that we have the slope and the y-intercept, we can write the equation for this table: $y = 1x + 5$, or $y = x + 5$. Looking back at the table, we can see that each value of y is 5 greater than its x value, so our equation is correct.

. . . .

TIP: If $x = 0$ in the input/output table, its y value is the y-intercept of the equation, because the y-intercept of an equation is the value of y when $x = 0$. In the previous example, the first row of the table shows that $y = 5$ when $x = 0$, and 5 is the y-intercept of the equation.

. . . .

We can find the equation used to make the following table in the same way.

Example

x	y
-2	-7
-1	-4
1	2
2	5
3	8

First, find the slope. Take the difference between the first two y values and divide it by the difference between the first two x values: $\frac{-4 - (-7)}{-1 - (-2)} = \frac{-4 + 7}{-1 + 2} = \frac{3}{1} = 3$. The slope of the line is 3. Next, find the y-intercept using the equation $y = mx + b$, the slope of the line, and one row of values from the table. Using the first row, we find that:

$$-7 = 3(-2) + b$$
$$-7 = -6 + b$$
$$-1 = b$$

The y-intercept of the line is –1, making the equation $y = 3x - 1$.

TIP: To find the slope and y-intercept using an input/output table, choose the rows of the table that contain the easiest numbers with which to work. The integers that are closest to zero are often the easiest values to add, subtract, multiply, and divide.

Practice 1

Find the equation that was used to build each table.

1.

x	y
1	5
2	10
3	15
4	20
5	25

2.

x	y
−2	4
−1	3
0	2
1	1
2	0

3.

x	y
2	−1
4	5
6	11
8	17
10	23

4.

x	y
−4	1
−2	2
2	4
4	5
6	6

FINDING THE MISSING VALUE IN A TABLE

You might have noticed we only need two rows from a table to find the equation of the table. Sometimes, we are asked to find a missing value from a table. We start by finding the equation of the table, and then we use that equation to find the missing value.

Example

x	y
1	3
2	−4
3	−11
4	−18
5	?

Using the first two rows of the table, we can find that the slope is $\frac{-4-3}{2-1} = \frac{-7}{1}$ = −7. Using the first row of the table, we can find the y-intercept of the equation:

$3 = -7(1) + b$
$3 = -7 + b$
$10 = b$

The equation for this table is $y = -7x + 10$. We are looking for the value of y when $x = 5$, so substitute 5 for x in the equation and solve for y:

$y = -7(5) + 10$
$y = -35 + 10$
$y = -25$

When $x = 5$, $y = -25$, so the missing value in the table is −25.

TIP: The missing value in a table might be an *x* value. After finding the equation for the table, substitute the *y* value that is in the same row as the missing *x* value into the equation and solve for *x*.

Practice 2

Find the missing value in each table.

1.

x	y
1	−8
2	−7
3	−6
4	−5
5	?

2.

x	y
−2	21
−1	13
0	?
1	−3
2	−11

3.

x	y
−2	−23
−1	−19
?	−11
3	−3
5	5

4.

x	y
−16	3
−8	1
8	−3
16	−5
24	?

ANSWERS

Practice 1

1. First, find the slope. Take the difference between the first two y values in the table and divide it by the difference between the first two x values in the table: $\frac{10-5}{2-1} = \frac{5}{1} = 5$. The slope of the line is 5. Next, find the y-intercept using the equation $y = mx + b$. Substitute 5, the slope of the line, for m. Substitute the values of x and y from the first row of the table, and solve for b, the y-intercept:

$$5 = 5(1) + b$$
$$5 = 5 + b$$
$$0 = b$$

The slope of the line is 5 and the y-intercept is 0. The equation of this line is $y = 5x + 0$, or $y = 5x$.

2. First, find the slope. Take the difference between the first two y values in the table and divide it by the difference between the first two x values in the table: $\frac{3-4}{-1-(-2)} = \frac{-1}{-1+2} = \frac{-1}{1} = -1$. The slope of the line is −1. Next, find the y-intercept using the equation $y = mx + b$. Substitute −1, the slope of the line, for m. Substitute the values of x and y from the third row of the table, and solve for b, the y-intercept:

$$2 = -1(0) + b$$
$$2 = 0 + b$$
$$2 = b$$

The slope of the line is −1 and the y-intercept is 2. The equation of this line is $y = -1x + 2$, or $y = -x + 2$.

3. First, find the slope. Take the difference between the first two y values in the table and divide it by the difference between the first two x values in the table: $\frac{5-(-1)}{4-2} = \frac{5+1}{2} = \frac{6}{2} = 3$. The slope of the line is 3. Next, find the y-intercept using the equation $y = mx + b$. Substitute 3, the slope of the line, for m. Substitute the values of x and y from the first row of the table, and solve for b, the y-intercept:

$$-1 = 3(2) + b$$
$$-1 = 6 + b$$
$$-7 = b$$

The slope of the line is 3 and the y-intercept is –7. The equation of this line is $y = 3x - 7$.

4. First, find the slope. Take the difference between the first two y values in the table and divide it by the difference between the first two x values in the table: $\frac{2-1}{-2-(-4)} = \frac{1}{-2+4} = \frac{1}{2}$. The slope of the line is $\frac{1}{2}$. Next, find the y-intercept using the equation $y = mx + b$. Substitute $\frac{1}{2}$, the slope of the line, for m. Substitute the values of x and y from the third row of the table, and solve for b, the y-intercept:

$$4 = \tfrac{1}{2}(2) + b$$
$$4 = 1 + b$$
$$3 = b$$

The slope of the line is $\frac{1}{2}$ and the y-intercept is 3. The equation of this line is $y = \frac{1}{2}x + 3$.

Practice 2

1. First, find the equation that was used to build the table. Use the first two rows of the table to find the slope: $\frac{-7-(-8)}{2-1} = \frac{-7+8}{1} = \frac{1}{1} = 1$. Use the first row of the table and the equation $y = mx + b$ to find the y-intercept of the equation. Remember: m is the slope and b is the y-intercept:

$$-8 = 1(1) + b$$
$$-8 = 1 + b$$
$$-9 = b$$

The equation for this table is $y = x - 9$. To find the value of y when $x = 5$, substitute 5 for x in the equation and solve for y:

$$y = (5) - 9$$
$$y = 5 - 9$$
$$y = -4$$

The missing value in the table is –4.

2. First, find the equation that was used to build the table. Use the first two rows of the table to find the slope: $\frac{(13-21)}{(-1-(-2))} = \frac{-8}{(-1+2)} = \frac{-8}{1} = -8$. Use the fourth row of the table and the equation $y = mx + b$ to find the y-intercept of the equation:

$$-3 = -8(1) + b$$
$$-3 = -8 + b$$
$$5 = b$$

The equation for this table is $y = -8x + 5$. To find the value of y when $x = 0$, substitute 0 for x in the equation and solve for y:

$$y = -8(0) + 5$$
$$y = 0 + 5$$
$$y = 5$$

The missing value in the table is 5.

3. First, find the equation that was used to build the table. Use the last two rows of the table to find the slope: $\frac{5-(-3)}{5-3} = \frac{5+3}{2} = \frac{8}{2} = 4$. Use the fourth row of the table and the equation $y = mx + b$ to find the y-intercept of the equation:

$$-3 = 4(3) + b$$
$$-3 = 12 + b$$
$$-15 = b$$

The equation for this table is $y = 4x - 15$. To find the value of x when $y = -11$, substitute -11 for y in the equation and solve for x:

$$-11 = 4x - 15$$
$$4 = 4x$$
$$1 = x$$

The missing value in the table is 1.

4. First, find the equation that was used to build the table. Use the first two rows of the table to find the slope: $\frac{1-3}{-8-(-16)} = \frac{-2}{-8+16} = \frac{-2}{8} = -\frac{1}{4}$. Use the second row of the table and the equation $y = mx + b$ to find the y-intercept of the equation:

$$1 = -\frac{1}{4}(-8) + b$$
$$1 = 2 + b$$
$$-1 = b$$

The equation for this table is $y = -\frac{1}{4}x - 1$. To find the value of y when $x = 24$, substitute 24 for x in the equation and solve for y:

$$y = -\frac{1}{4}(24) - 1$$
$$y = -6 - 1$$
$$y = -7$$

The missing value in the table is -7.

graphing equations

Mathematical genius and artistic genius touch one another.
—GÖSTA MITTAG-LEFFLER (1846–1927)
SWEDISH MATHEMATICIAN

In this lesson, you'll learn about the coordinate plane and how to graph the equation of a line.

THE EQUATION OF A LINE is a linear equation. A **linear equation** is an equation that can contain constants and variables, and the exponent of the variables is 1. We can graph linear equations on the coordinate plane. The **coordinate plane** is a two-dimensional surface with an x-axis and a y-axis. The **x-axis** is a horizontal line along which $y = 0$, and the **y-axis** is a vertical line along which $x = 0$. The coordinate plane is shown on the following page.

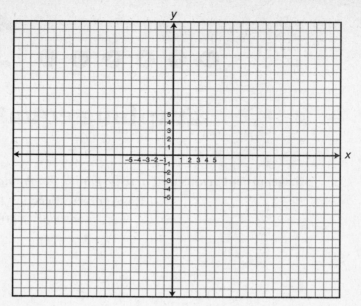

The first few lines to each side of the axes are labeled. The x values increase as we move to the right, and the y values increase as we move up. Points in the upper right corner, or quadrant I, of the plane have positive x values and positive y values. Points in the upper left corner, or quadrant II, have negative x values and positive y values. Points in the lower left corner, quadrant III, have negative x values and negative y values, and points in the lower right corner, quadrant IV, have positive x values and negative y values.

We plot points on the coordinate plane according to their x and y values. These values are called an **ordered pair** or a **coordinate pair**. In an ordered pair, the x value is listed first, and then the y value, like this: (4,2). The x value of this point is 4 and the y value of the point is 2.

Each row in the input/output tables from the preceding lesson represents an ordered pair. Following is the input/output table from the equation $y = 3x - 1$.

x	y
–2	–7
–1	–4
1	2
2	5
3	8

We can plot each of these points on the coordinate plane. The first row shows that x is –2 and y is –7. This is the ordered pair (–2,–7). To plot this point, start at the origin of the coordinate plane. The origin is the place where the *x*-axis and *y*-axis cross, where $x = 0$ and $y = 0$. Because the *x* value of this point is –2, move two units to the left. The *y* value of the point is –7, so move seven units down. Place a dot where $x = -2$ and $y = -7$, and label it (–2,–7). Do the same for all five rows of the table. The points in rows three, four, and five of the table will be plotted in quadrant I, because the *x* and *y* values are both positive.

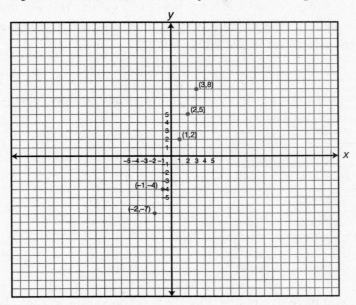

Finally, connect the dots with a solid line, and put arrows on both ends of the line to show that the line continues in both directions. Label the line with its equation, and you have finished. That's all there is to graphing the equation of a line!

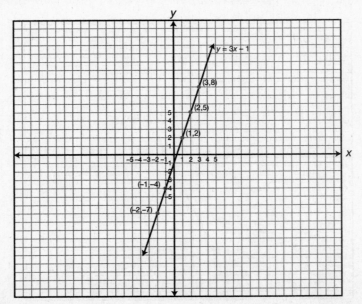

How do we graph a line if we do not have an input/output table? We create our own input/output table. Choose a few x values, and use the equation of the line to find their y values.

···

TIP: When you are making an input/output table, it is helpful to choose a few positive x values and a few negative x values. This will give you points on both sides of the y-axis. Also, include $x = 0$ in your table, so that the y-intercept is plotted.

···

To graph the line $y = 2x - 6$, start with an input/output table:

	$y = 2x - 6$	
x		y
−2	2(−2) − 6	−10
−1	2(−1) − 6	−8
0	2(0) − 6	−6
1	2(1) − 6	−4
2	2(2) − 6	−2

Now, you have five points. Plot these points on the graph. The first two points, (−2,−10) and (−1,−8), will be in quadrant III, since the x and y values are both negative. The last two points, (1,−4) and (2,−2) will be in quadrant IV. Connect the dots to form the line $y = 2x - 6$.

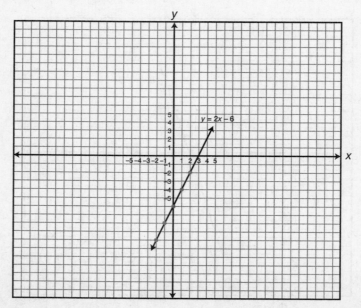

TIP: If the slope of the equation is a fraction, choose *x* values for your input/output table that are multiples of the denominator of the fraction. The *y* values will be integers, which are easier to graph than fractions.

Practice

Graph each equation.

1. $y = x + 4$

2. $y = -3x + 1$

3. $y = \frac{1}{2}x - 3$

4. $y = \frac{3}{4}x + 6$

ANSWERS

Practice

1. To graph the line $y = x + 4$, start with an input/output table. Choose two negative numbers, two positive numbers, and 0 as the *x*-values:

x	$y = x + 4$	y
-2	(-2) + 4	2
-1	(-1) + 4	3
0	(0) + 4	4
1	(1) + 4	5
2	(2) + 4	6

Plot these points on the graph. The first two points, (-2,2) and (-1,3), will be in quadrant II, because the *x* values are negative and the *y* values are positive. The last two points, (1,5) and (2,6) will be in quadrant I, because the x and y values are positive. Connect the dots to form the line $y = x + 4$.

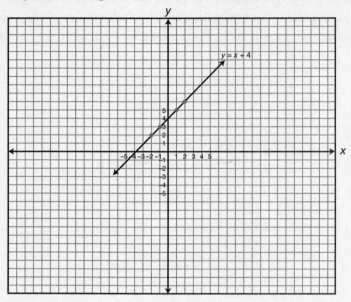

2. To graph the line $y = -3x + 1$, start with an input/output table. Choose two negative numbers, two positive numbers, and 0 as the x-values:

$y = -3x + 1$		
x		y
−2	−3(−2) + 1	7
−1	−3(−1) + 1	4
0	−3(0) + 1	1
1	−3(1) + 1	−2
2	−3(2) + 1	−5

Plot these points on the graph. The first two points, (−2,7) and (−1,4), will be in quadrant II, because the x values are negative and the y values are positive. The last two points, (1,−2) and (2,−5), will be in quadrant IV, because the x values are positive and the y values are negative. Connect the dots to form the line $y = -3x + 1$.

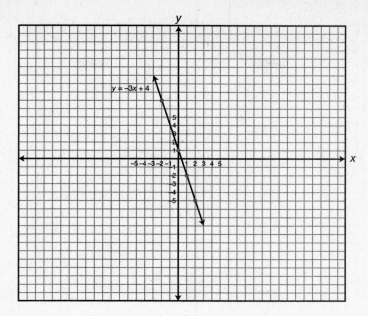

3. To graph the line $y = \frac{1}{2}x - 3$, start with an input/output table. Choose x values that are even, so that the y values will be integers:

$y = \frac{1}{2}x - 3$		
x		y
-4	$\frac{1}{2}(-4) - 3$	-5
-2	$\frac{1}{2}(-2) - 3$	-4
0	$\frac{1}{2}(0) - 3$	-3
2	$\frac{1}{2}(2) - 3$	-2
4	$\frac{1}{2}(4) - 3$	-1

Plot these points on the graph. The first two points, $(-4,-5)$ and $(-2,-4)$, will be in quadrant III, because the x and y values are negative. The last two points, $(2,-2)$ and $(4,-1)$, will be in quadrant IV, because the x values are positive and the y values are negative. Connect the dots to form the line $y = \frac{1}{2}x - 3$.

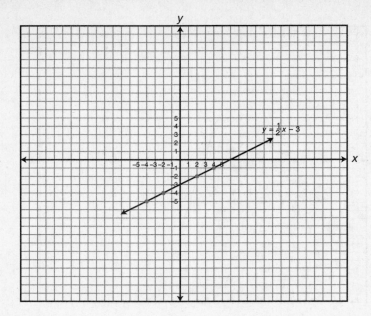

4. To graph the line $y = \frac{3}{4}x + 6$, start with an input/output table. Choose x-values that are multiples of 4, so that the y values will be integers:

$y = \frac{3}{4}x + 6$		
x		y
-8	$\frac{3}{4}(-8) + 6$	0
-4	$\frac{3}{4}(-4) + 6$	3
0	$\frac{3}{4}(0) + 6$	6
4	$\frac{3}{4}(4) + 6$	9
8	$\frac{3}{4}(8) + 6$	12

Plot these points on the graph. The second point, $(-4,3)$, will be in quadrant II, because the x value is negative and the y value is positive. The last two points, $(4,9)$ and $(8,12)$, will be in quadrant I, because the x and y values are positive. Connect the dots to form the line $y = \frac{3}{4}x + 6$.

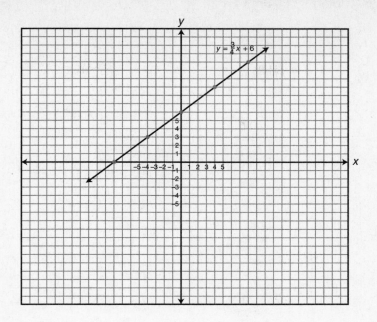

finding equations from graphs

The simplest schoolboy is now familiar with facts for which
Archimedes would have sacrificed his life.
—ERNEST RENAN (1823–1892)
FRENCH PHILOSOPHER

In this lesson, you'll learn how to find the equation of a line from the graph of
a line, and how to determine if an ordered pair is on a line.

NOW THAT WE KNOW how to graph the equation of a line, we will go in the
other direction: Given the graph of a line, we will find its equation. Just as with
input/output tables and graphing, we start with the slope. Because the slope of
a line is the change in the *y* values between two points divided by the change
in the *x* values of those points, we can find the slope by looking at the line and
choosing any two points on it. Look at the following graph.

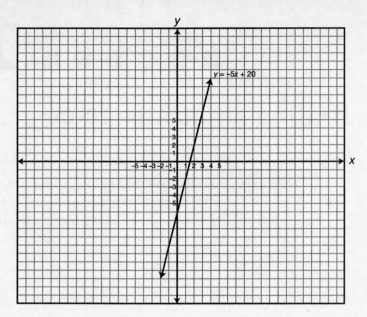

Although we can find the slope using any two points, choose points for which x and y are integers. On this graph, when $y = 2$, it is difficult to see the value of x. Its value is somewhere between 3 and 4, but we cannot be certain. However, it is easy to see that when $x = 3$, $y = 5$, and when $x = 4$, $y = 0$. We'll use these points to find the slope of the line: $\frac{0-5}{4-3} = \frac{-5}{1} = -5$. The slope of the line is -5. The next step might seem familiar: To find the y-intercept, use the slope, a point on the graph, and the equation $y = mx + b$. Let's use the point (4,0):

$0 = -5(4) + b$
$0 = -20 + b$
$20 = b$

The y-intercept of the graph is 20, which means that this is the graph of the equation $y = -5x + 20$.

....

TIP: Sometimes, you can look right on the graph to find the y-intercept. If you can see where the line crosses the y-axis, you will have the y-intercept of the equation without having to perform any calculations.

....

Example

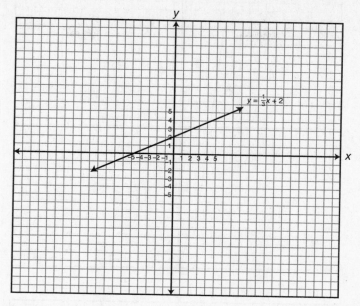

    To find the equation of the line graphed here, we begin again with the slope. Looking at the graph, when $x = 3$, $y = 3$, and when $x = 6$, $y = 4$. We'll use the points $(3,3)$ and $(6,4)$ to find the slope: $\frac{4-3}{6-3} = \frac{1}{3}$. We could calculate the y-intercept, but look closely at the graph. The line crosses the y-axis where $y = 2$, which means that 2 is the y-intercept. This is the graph of the equation $y = \frac{1}{3}x + 2$.

Practice 1

Find the equation of the line on each graph.

1.

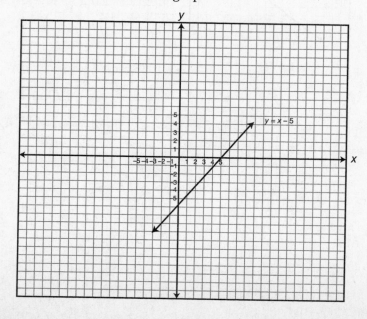

2.

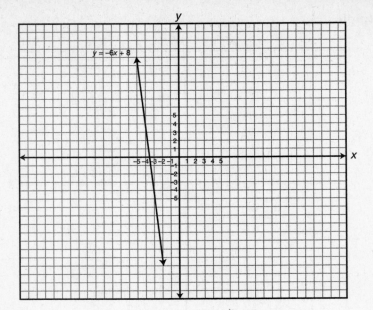

3.

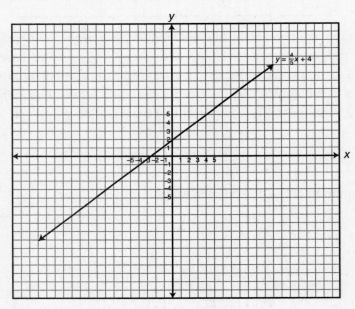

4.

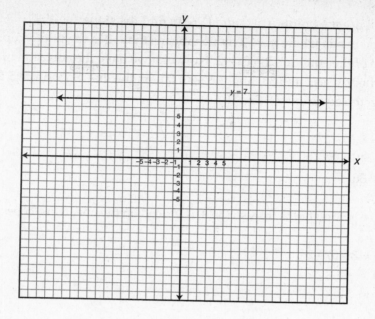

ORDERED PAIRS

If we are given a point, or ordered pair, and the graph of a line, we can determine if that ordered pair is on the line. We might be able to tell just by looking at the graph. If not, we can find the equation of the line, and then substitute the values of the ordered pair into the equation to see if they hold true.

Are the ordered pairs (5,38) and (−6,−46) on the following line graphed? We cannot tell just by looking, so we must find the equation of the line.

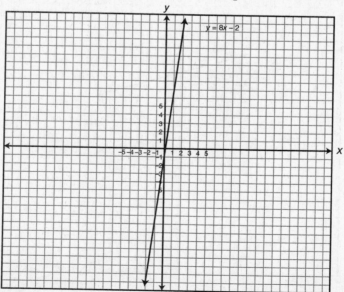

We'll use the points $(0,-2)$ and $(1,6)$ to find the slope: $\frac{6-(-2)}{1-0} = \frac{8}{1} = 8$. The line crosses the y-axis where $y = -2$, which means that -2 is the y-intercept. This is the graph of the equation $y = 8x - 2$. Now that we have the equation of the line, we will check to see if $(5,38)$ and $(-6,-46)$ fall on the line. Substitute 5 for x and 38 for y in the equation $y = 8x - 2$:

$$38 = 8(5) - 2 ?$$
$$38 = 40 - 2 ?$$
$$38 = 38$$

The equation holds true, so $(5,38)$ is indeed on the line. Now, check $(-6,-46)$:

$$-46 = 8(-6) - 2 ?$$
$$-46 = -48 - 2 ?$$
$$-46 \neq -50$$

-46 does not equal -50, so the point $(-6,-46)$ is not on the line $y = 8x - 2$.

Practice 2

Determine whether the points $(20,2)$ and $(-20,10)$ fall on each of the following graphed lines below.

 1.

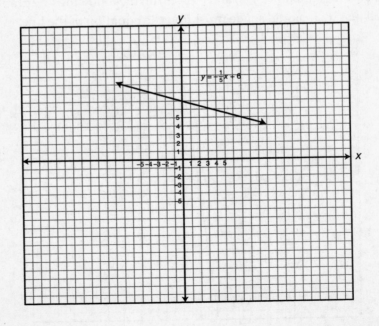

2.

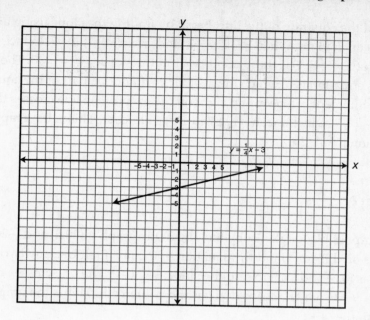

$y = \frac{1}{4}x - 3$

ANSWERS

Practice 1

1. To find the equation of the line, begin by finding the slope using any two points on the line. When $x = 1$, $y = -4$, and when $x = 2$, $y = -3$. Use these points to find the slope: $\frac{-3-(-4)}{2-1} = \frac{-3+4}{1} = \frac{1}{1} = 1$. The y-intercept can be found right on the graph. The line crosses the y-axis where $y = -5$, which means that -5 is the y-intercept. This is the graph of the equation $y = x - 5$.

2. To find the equation of the line, begin by finding the slope using any two points on the line. When $x = -4$, $y = 6$, and when $x = -3$, $y = 0$. Use these points to find the slope: $\frac{0-6}{-3-(-4)} = \frac{-6}{-3+4} = \frac{-6}{1} = -6$. To find the y-intercept, use the slope, any point on the graph, and the equation $y = mx + b$. Using the point $(-3,0)$:

$$0 = -6(-3) + b$$
$$0 = 18 + b$$
$$-18 = b$$

This is the graph of the equation $y = -6x - 18$.

3. To find the equation of the line, begin by finding the slope using any two points on the line. When $x = -5$, $y = 0$, and when $x = 0$, $y = 4$. Use these points to find the slope: $\frac{4-0}{0-(-5)} = \frac{4}{0+5} = \frac{4}{5}$. The y-intercept can be found right on the graph. The line crosses the y-axis where $y = 4$, which means that 4 is the y-intercept. This is the graph of the equation $y = \frac{4}{5}x + 4$.

4. To find the equation of the line, begin by finding the slope using any two points on the line. When $x = 0$, $y = 7$, and when $x = 1$, $y = 7$. In fact, for any value of x, $y = 7$. Use these points to find the slope: $\frac{7-7}{1-0} = \frac{0}{1} = 0$. The slope of this equation is 0, which means that x will not be part of the equation. The y-intercept can be found right on the graph. The line crosses the y-axis where $y = 7$, which means that 7 is the y-intercept. This is the graph of the equation $y = 7$.

Practice 2

1. It is impossible to tell just by looking at the graph if (20,2) and (–20,10) are points on this line. Find the equation of the line first. Use any two points, such as (0,6) and (5,5) to find the slope: $\frac{5-6}{5-0} = -\frac{1}{5}$. The line crosses the y-axis where $y = 6$, which means that 6 is the y-intercept. This is the graph of the equation $-\frac{1}{5}x + 6$.

Check to see if (20,2) falls on the line by substituting 20 for x and 2 for y:

$2 = -\frac{1}{5}(20) + 6$?

$2 = -4 + 6$?

$2 = 2$

The equation holds true, so (20, 2) is on the line $y = -\frac{1}{5}x + 6$. Check (–20, 10):

$10 = -\frac{1}{5}(-20) + 6$?

$10 = 4 + 6$?

$10 = 10$

The equation holds true again, so (–20,10) is also on the line $y = -\frac{1}{5}x + 6$.

2. It is impossible to tell just by looking at the graph if (20, 2) and (–20, 10) are points on this line. Find the equation of the line first. Use any two points, such as (0,–3) and (4,–2) to find the slope: $\frac{-2-(-3)}{4-0} = \frac{-2+3}{4} = \frac{1}{4}$. The line crosses the y-axis where $y = -3$, which means that –3 is the y-intercept. This is the graph of the equation $\frac{1}{4}x - 3$.

Check to see if (20, 2) falls on the line by substituting 20 for x and 2 for y:

$2 = \frac{1}{4}(20) - 3$?

$2 = 5 - 3$?

$2 = 2$

The equation holds true, so (20,2) is on the line $y = \frac{1}{4}x - 3$. Check (–20,10):

$10 = \frac{1}{4}(-20) - 3$?

$10 = -5 - 3$?

$10 \neq -8$

The equation does not hold true, so (–20,10) is not on the line $y = \frac{1}{4}x - 3$.

distance

Whenever you can, count.
—SIR FRANCIS GALTON (1822–1911)
ENGLISH GENETICIST AND STATISTICIAN

In this lesson, you'll learn how to find the distance between two points on a graph by counting, using the Pythagorean theorem, and using the distance formula.

A LINE ON THE COORDINATE PLANE continues forever in both directions, but we can find the distance between two points on the line. When the points are on a horizontal line, such as $y = 3$, we can simply count the units from one point to the other.

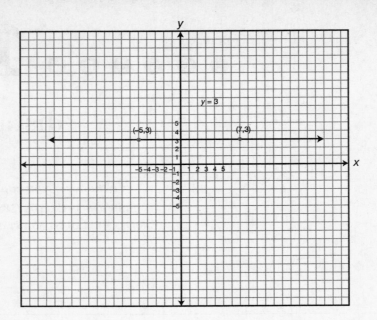

What is the distance from point (–5,3) to point (7,3)? We can count the unit boxes between the two points. There are 12 units from (–5,3) to (7,3), which means that the distance between the points is 12 units.

··

TIP: If the y values of two points are the same, the distance between the two points is equal to the difference between their x values. If the x values of two points are the same, the distance between the two points is equal to the difference between their y values.

··

The distance between (5,4) and (5,–10) is 14 units, because the x values of the points are the same, and $4 - (-10) = 14$ units.

Often though, we need to find the distance between two points that are not on a horizontal or vertical line. For instance, how can we find the distance between two points on the line $y = x$?

PYTHAGOREAN THEOREM

The **Pythagorean theorem** describes the relationship between the sides of a right triangle. It states that the sum of the squares of the bases of the triangle is equal to the square of the hypotenuse of the triangle: $a^2 + b^2 = c^2$. What is a triangle formula doing in a lesson about the distance between two points on a line? Drawing a triangle on a graph can help us find the distance between two points.

The following graph shows the line $y = \frac{4}{3}x + 2$, which contains the points (3,6) and (6,10).

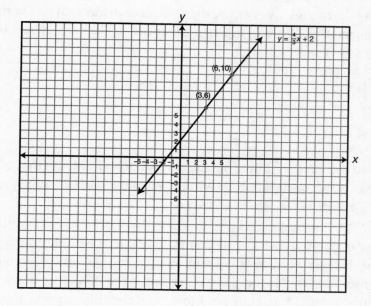

We can't find the distance between these points just by counting. However, we can draw a vertical line down from (6,10) and a horizontal line from (3,6). These points meet at (6,6) and form a right triangle.

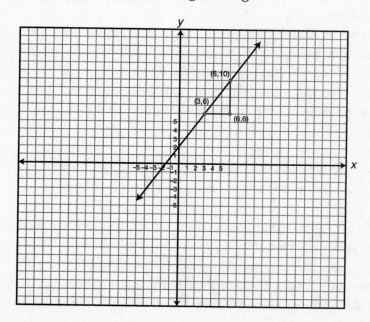

The bases of the triangle are the line segment from (3,6) to (6,6) and the line segment from (6,6) to (6,10). Because a horizontal line connects (3,6) to (6,6), we can find its length just by counting. The length of this line is 3 units. In the same way, the distance from (6,6) to (6,10) is 4 units. Now that we know the length of each base of the triangle, we can use the Pythagorean theorem to find the length of the hypotenuse of the triangle. In the formula $a^2 + b^2 = c^2$, a and b are the bases. Substitute 3 for a and 4 for b:

$$3^2 + 4^2 = c^2$$
$$9 + 16 = c^2$$
$$25 = c^2$$

To find the value of c, take the square root of both sides of the equation. A distance can never be negative, so we only need the positive square root of 25, which is 5. The distance between (3,6) and (6,10) is 5 units.

You might be thinking, "I don't want to draw a triangle every time I need to find the distance between two points." Well, let's look at exactly how we found the distance between (3,6) and (6,10). First, we added a point to the graph to form a triangle. The point represented the difference in the x values between the points (3 units, from 3 to 6) and the difference in the y values between the points (4 units, from 6 to 10). Then, we squared those differences, added them, and took the square root. To make finding distance easier, we can write these steps as a formula: the distance formula.

The distance formula states that $D = \sqrt{(x_2 - x_1)^2 + (y_2 - y_1)^2}$. That doesn't look much easier, so let's break the formula down into pieces. We want to find the distance from point 1, (3,6) and point 2, (6,10). To find the difference between the x values of the points, we subtract the first x value, which we can write as x_1, from the second x value, which we can write as x_2. To find the difference between the y values of the points, we subtract the first y value, y_1, from the second y value, y_2.

Difference between the x values: $(x_2 - x_1) = (6 - 3) = 3$
Difference between the y values: $(y_2 - y_1) = (10 - 6) = 4$

Now that we have the differences between each value, we square them and add them. This is the part of the formula that comes from the Pythagorean theorem. The square of 3 is 9 and the square of 4 is 16, and $9 + 16 = 25$.

Finally, this sum is equal to the square of the distance between the two points, so to find the distance, we must take the square root of the sum: $\sqrt{25} = 5$.

Now, not only do you know how to use the distance formula, you know where it comes from!

> **Example**
> Find the distance between (–2,4) and (3,16).

We will call (–2,4) point 1 and (3, 16) point 2. Substitute these values into the distance formula: $D = \sqrt{(x_2 - x_1)^2 + (y_2 - y_1)^2}$. Because (–2,4) is the first point, x_1 is –2 and y_1 is 4. The value of x_2 is 3 and the value of y_2 is 16:

$$D = \sqrt{(3-(-2))^2 + (16-4)^2}$$

Remember the order of operations: parentheses come before exponents, so perform the subtraction first:

$$D = \sqrt{(5)^2 + (12)^2}$$

Exponents come before addition, so square 5 and 12 next:

$$D = \sqrt{25 + 144}$$
$$D = \sqrt{169}$$
$$D = 13 \text{ units}$$

The distance between (–2,4) and (3,16) is 13 units.

RADICAL DISTANCES

Sometimes, the distance between two points is not a whole number, and we are left with a radical. Let's find the distance between (1,2) and (7,12). Enter the x and y values of each point into the distance formula:

$$D = \sqrt{(7-1)^2 + (12-2)^2}$$
$$D = \sqrt{(6)^2 + (10)^2}$$
$$D = \sqrt{36 + 100}$$
$$D = \sqrt{136}$$

The number 136 is not a perfect square. When the distance formula leaves you with a radical like this, the only way to simplify it is to factor out a perfect square. Check to see if the radicand is divisible by 4, 9, 25, or some other square. The number 136 is divisible by 4, which means that $\sqrt{136}$ is divisible by $\sqrt{4}$:

$\sqrt{136} = \sqrt{4}\sqrt{34}$, because we can factor a radicand into two radicands in the same way that we factor a whole number into two whole numbers. The square root of 4 is 2, so $\sqrt{4}\sqrt{34}$ is equal to $2\sqrt{34}$. The distance between (1,2) and (7,12) is $2\sqrt{34}$ units.

Let's look at one last example: the distance between (2,1) and (–2,10). Enter the x and y values of each point into the distance formula:

$$D = \sqrt{((-2)-2)^2 + (10-1)^2}$$
$$D = \sqrt{(-4)^2 + (9)^2}$$
$$D = \sqrt{16+81}$$
$$D = \sqrt{97}$$

The number 97 is not divisible by any whole number perfect square (4, 25, 36, 49, 64, or 81), so $\sqrt{97}$ cannot be simplified. The distance between (2,1) and (–2,10) is $\sqrt{97}$ units.

Practice

Find the distance between each pair of points.

1. (2,8) and (8,16)

2. (–5,1) and (4,13)

3. (3,4) and (8,3)

4. (–7,–1) and (–1,–3)

5. (–3,–5) and (5,–13)

ANSWERS

Practice

1. Use the distance formula to find the distance between (2,8) and (8,16).
   Because (2,8) is the first point, x_1 is 2 and y_1 is 8. (8,16) is the second point, so x_2 is 8 and y_2 is 16:

   $$D = \sqrt{(x_2 - x_1)^2 + (y_2 - y_1)^2}$$
   $$D = \sqrt{(8-2)^2 + (16-8)^2}$$
   $$D = \sqrt{(6)^2 + (8)^2}$$
   $$D = \sqrt{36 + 64}$$
   $$D = \sqrt{100}$$
   $$D = 10 \text{ units}$$

2. Use the distance formula to find the distance between (−5,1) and (4,13).
   Because (−5,1) is the first point, x_1 is −5 and y_1 is 1. (4,13) is the second point, so x_2 is 4 and y_2 is 13:

   $$D = \sqrt{(x_2 - x_1)^2 + (y_2 - y_1)^2}$$
   $$D = \sqrt{(4-(-5))^2 + (13-1)^2}$$
   $$D = \sqrt{(9)^2 + (12)^2}$$
   $$D = \sqrt{81 + 144}$$
   $$D = \sqrt{225}$$
   $$D = 15 \text{ units}$$

3. Use the distance formula to find the distance between (3,4) and (8,3).
   Because (3,4) is the first point, x_1 is 3 and y_1 is 4. (8,3) is the second point, so x_2 is 8 and y_2 is 3:

   $$D = \sqrt{(x_2 - x_1)^2 + (y_2 - y_1)^2}$$
   $$D = \sqrt{(8-3)^2 + (3-4)^2}$$
   $$D = \sqrt{(5)^2 + (-1)^2}$$
   $$D = \sqrt{25 + 1}$$
   $$D = \sqrt{26} \text{ units}$$

4. Use the distance formula to find the distance between $(-7,-1)$ and $(-1\,-3)$. Because $(-7,-1)$ is the first point, x_1 is -7 and y_1 is -1. $(-1,-3)$ is the second point, so x_2 is -1 and y_2 is -3:

$$D = \sqrt{(x_2 - x_1)^2 + (y_2 - y_1)^2}$$
$$D = \sqrt{(-1 - (-7))^2 + (-3 - (-1))^2}$$
$$D = \sqrt{(6)^2 + (-2)^2}$$
$$D = \sqrt{36 + 4}$$
$$D = \sqrt{40}$$

The number 40 is divisible by 4, which is a perfect square. Because $40 = (4)(10)$, $\sqrt{40} = \sqrt{4}\,\sqrt{10} = 2\sqrt{10}$ units.

5. Use the distance formula to find the distance between $(-3,-5)$ and $(5,-13)$. Because $(-3,-5)$ is the first point, x_1 is -3 and y_1 is -5. $(5,-13)$ is the second point, so x_2 is 5 and y_2 is -13:

$$D = \sqrt{(x_2 - x_1)^2 + (y_2 - y_1)^2}$$
$$D = \sqrt{(5 - (-3))^2 + (-13 - (-5))^2}$$
$$D = \sqrt{(8)^2 + (-8)^2}$$
$$D = \sqrt{64 + 64}$$
$$D = \sqrt{128}$$

The number 128 is divisible by 64, which is a perfect square. Because $128 = (64)(2)$, $\sqrt{128} = \sqrt{64}\sqrt{2} = 8\sqrt{2}$ units.

functions, domain, and range

Mathematics is an escape from reality.
—STANISLAW ULAM (1909–1984)
POLISH MATHEMATICIAN

In this lesson, you'll learn how to determine if an equation is a function, and how to find the domain and range of a function.

ALMOST EVERY LINE we have seen in the last few lessons has been a **function**. An equation is a function if every x value has no more than one y value. For instance, the equation $y = 2x$ is a function, because there is no value of x that could result in two different y values. When an equation is a function, we can replace y with $f(x)$, which is read as "f of x." If you see an equation written as $f(x) = 2x$, you are being told that y is a function of x, and that the equation is a function.

The equation $x = 5$ is not a function. When x is 5, y has many different values. Vertical lines are not functions. They are the only type of line that is not a function.

What about the equation $y = x^2$? Positive and negative values of x result in the same y value, but that is just fine. A function can have y values that each have more than one x value, but a function *cannot* have x values that each have more than one y value. There is no number that can be substituted for x that results in two different y values, so $y = x^2$ is a function.

What about the equation $y^2 = x$? In this case, two different y values, such as 2 and –2, result in the same x value, 4, so $y^2 = x$ is not a function. We must always be careful with equations that have even exponents. If we take the square root of both sides of the equation $y^2 = x$, we get $y = \sqrt{x}$, which *is* a function. x cannot be negative, because we cannot find the square root of a negative number. Because x must be positive, y must be positive. This means that, unlike $y^2 = x$, there is no x value having two y values.

....

TIP: When you are trying to decide whether an equation is a function, always ask: Is there a value of x having two y values? If so, the equation is not a function.

....

Practice 1

Identify whether each equation is a function.

1. $y = 100x$

2. $y = 5$

3. $x = 4$

4. $x = 4y$

5. $|y| = x$

VERTICAL LINE TEST

When we can see the graph of an equation, we can easily identify whether the equation is a function by using the **vertical line test**. If a vertical line can be drawn anywhere through the graph of an equation, such that the line crosses the graph more than once, then the equation is not a function. Why? Because a vertical line represents a single x value, and if a vertical line crosses a graph more than once, then there is more than one y value for that x value.

Look at the following graph. We do not know what equation is shown, but we know that it is a function, because there is no place on the graph where a vertical line will cross the graph more than once.

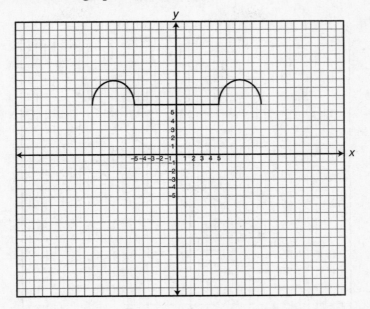

Even if there are many places on a graph that pass the vertical line test, if there is even one point for which the vertical line test fails, then the equation is not a function. The following graph, a circle, is not a function, because there are many x values that have two y values. The dark line drawn where $x = 5$ shows that the graph fails the vertical line test. The line crosses the circle in two places.

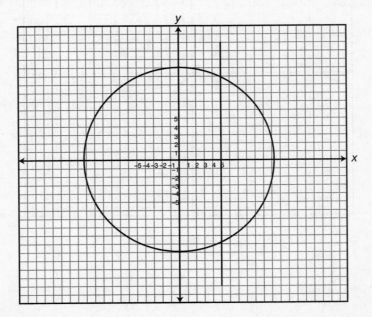

Practice 2

Use the vertical line test to identify whether each graph is a function.

1.

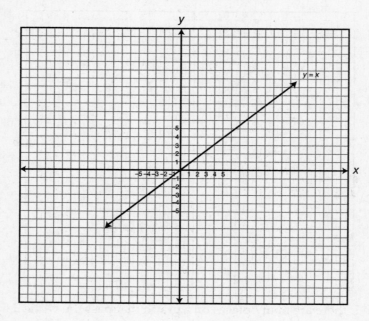

2.

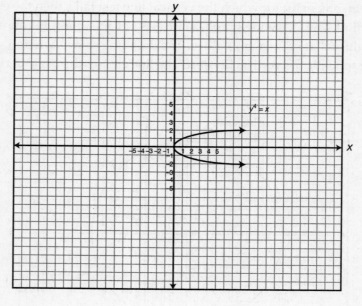

3.

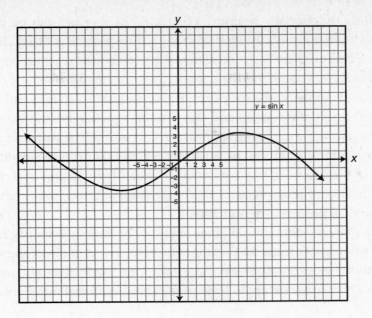

4.

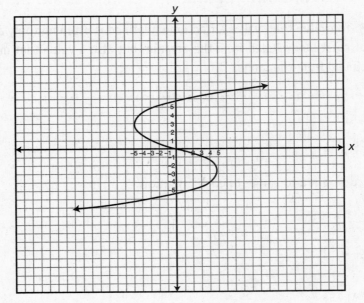

DOMAIN AND RANGE

Earlier, we looked at the equation $y = \sqrt{x}$, and stated that x could not be negative, because we cannot take the square root of a negative number. However, for the equation $y = x$, we could make x any real number. For this equation, we say that the domain is all real numbers. The **domain** of an equation or function is all the values that can be substituted for x. In the equation $y = \sqrt{x}$, the domain is zero and all positive real numbers.

While the domain describes what kind of x values can be put into an equation, the **range** tells us what kind of y values we will get back. In the equation $y = x$, because x can be any real number, y can be any real number. The range of $y = x$ is all real numbers. The equation $y = \sqrt{x}$ has a domain of zero and all positive real numbers, and the possible values of y are also zero and all positive real numbers. So the domain and range of this function are the same as well.

The domain and range are not always the same. The equation $y = x^2$ has a domain of all real numbers, because any real number can be substituted for x. However, the square of any real number, including negative real numbers, is always positive. There will be no y values that are negative, so the range of $y = x^2$ is zero and all positive real numbers.

What are the domain and range of $y = \frac{1}{x}$? The denominator of a fraction can never be 0, which means that x cannot be zero. The domain of the equation is all real numbers except 0. When these values are put into the equation, we can get back any y value except 0. The range of $y = \frac{1}{x}$ is all real numbers except 0.

....

TIP: The value that makes a fraction undefined is often not only a value that must be excluded from the domain, but also a value that must be excluded from the range.

....

Practice 3

Find the domain and range of each equation.

1. $y = x - 12$

2. $y = x^3$

3. $y = |x|$

4. $y = \frac{5}{7-x}$

5. $y = x^2 + 3$

ANSWERS

Practice 1

1. Every value that can be substituted for x in the equation $y = 100x$ will give us one y value and one y value only, so $y = 100x$ is a function.
2. In the equation $y = 5$, the value of x does not matter. The value of y will always be 5, which means that for any x value, there is exactly one y value. Because there can never be more than one y value for any x value, $y = 5$ is a function.
3. In the equation $x = 4$, only one value can be substituted for x, and that is 4. Only 4 is equal to 4. When $x = 4$, y has many different values, so $x = 4$ is not a function.
4. The equation $x = 4y$ is a little easier to understand when it is rewritten in $y =$ form. Divide both sides of the equation by 4: $x = 4y$ is the same as $y = \frac{1}{4}x$. Every value that can be substituted for x in this equation will give us one y value and one y value only, so $x = 4y$ is a function.
5. The equation $|y| = x$ means that the absolute value of y is equal to x. When the absolute value of a number is taken, any negative sign is removed from the number. The absolute value of 3 and the absolute value of -3 are both 3. This means that every positive value of x has two y values, a positive value and a negative value, so the equation $|y| = x$ is not a function.

Practice 2

1. The following graph passes the vertical line test, because a vertical line can be drawn anywhere and it will cross the graph in no more than one place. This graph is a function.

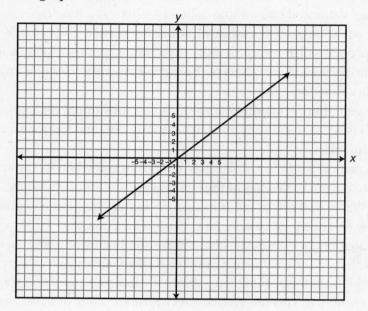

2. The following graph does not pass the vertical line test, because a vertical line can be drawn anywhere on the right side of the y-axis and it will cross the graph in two places. This graph is not a function.

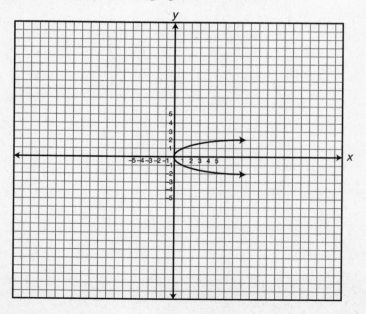

3. This graph passes the vertical line test, because a vertical line can be drawn anywhere and it will cross the graph in no more than one place. This graph is a function.

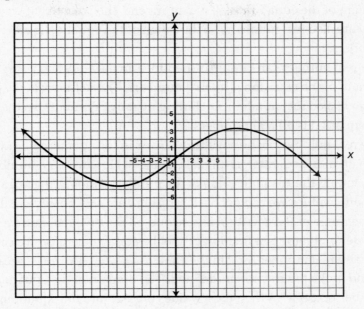

4. The following graph does not pass the vertical line test, because a vertical line can be drawn anywhere between $x = -4$ and $x = 4$ and it will cross the graph in three places. This graph is not a function.

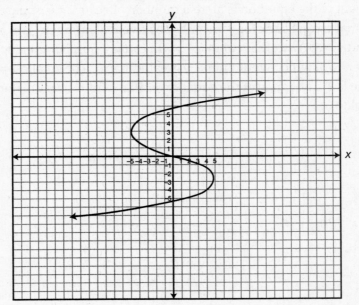

Practice 3

1. Any real number can be substituted for x in the equation $y = x - 12$, so the domain of the equation is all real numbers. The values that we get back for y are also real numbers, so the range of the equation is all real numbers.

2. Any real number can be substituted for x in the equation $y = x^3$, so the domain of the equation is all real numbers. The values that we get back for y are also real numbers, including negative numbers, so the range of the equation is all real numbers.

3. Any real number can be substituted for x in the equation $y = |x|$, so the domain of the equation is all real numbers. However, if a number is negative, taking the absolute value of the number removes the negative sign. The only values we can get back for y are zero and positive real numbers, so the range of the equation is zero and all positive real numbers.

4. The denominator of a fraction cannot be equal to zero, so we can substitute any real number for x in the equation $y = \frac{5}{7-x}$ except 7, because 7 would make the fraction undefined. The domain of the equation is all real numbers except 7. When these values are put into the equation, we can get back any y value except 0. The range of the equation is all real numbers except 0.

5. Any real number can be substituted for x in the equation $y = x^2 + 3$, so the domain of the equation is all real numbers. Because x is squared, the values that we get back for y are only positive real numbers. If $x = 0$, then $x^2 = 0$, and $y = 3$. Any other value for x will cause a number greater than 0 to be added to 3, which would make y greater than 3. The smallest possible y value is 3. The range of the equation is all real numbers greater than or equal to 3.

systems of equations— solving with substitution

In mathematics, you don't understand things.
You just get used to them.
—Johann von Neumann (1903–1957)
Hungarian-American mathematician

In this lesson, you'll learn how to solve systems of equations with two or three variables using substitution.

WE LEARNED HOW to solve an equation with one variable in Lesson 9. In that same lesson, we saw that we could not find the values of two variables with just one equation. The best we could do was to solve for one variable in terms of the other. For example, if $3x + y = 7$, we could subtract $3x$ from both sides of the equation to solve for y in terms of x:

$3x - 3x + y = 7 - 3x$
$y = 7 - 3x$

Once we had y in terms of x, there was nothing more we could do. In order to find the values of x and y, we need a second, related equation. A group of two or more equations for which the common variables in each equation have the same values is called a **system of equations**.

The equation $3x + y = 7$ and the equation $x - y = 13$ form a system of equations. The values of x and y in the first equation are also the values of x and y in the second equation. Now that we have two equations, we can find the values of both variables.

We have a couple of ways to solve a system of equations. In this lesson, we will learn how to solve them using substitution. In the next lesson, we will learn how to solve them using elimination.

When we solve a system of equations using substitution, we begin by writing one variable in terms of the other. In this example, we wrote y in terms of x: $y = 7 - 3x$. Next, we replace that variable in the other equation with its expression. In this example, we replace the y in the second equation, $x - y = 13$, with $7 - 3x$, because that is the value of y. The second equation is now $x - (7 - 3x) = 13$. Now we have a single equation with a single variable. Solve for the value of the variable:

$$x - (7 - 3x) = 13$$
$$x - 7 + 3x = 13$$
$$4x - 7 = 13$$
$$4x - 7 + 7 = 13 + 7$$
$$4x = 20$$
$$x = 5$$

We now have the value of one variable, x. We can replace x with its value in either equation to find the value of y. Let's use the second equation:

$$x - y = 13$$
$$5 - y = 13$$
$$-y = 8$$
$$y = -8$$

The solution to this system of equations is $x = 5$, $y = -8$. Substitute the value of each variable into both equations to check the answer:

$3x + y = 7$	$x - y = 13$
$3(5) + (-8) = 7$	$5 - (-8) = 13$
$15 - 8 = 7$	$5 + 8 = 13$
$7 = 7$	$13 = 13$

TIP: If a variable in one equation of a system of equations does not have a coefficient, begin by solving for that variable in terms of the other variable. By avoiding division, you may avoid working with fractions.

Example
$3y = 12x + 3$
$5x + y = 10$

Start by writing one variable in terms of the other. We could write x in terms of y using either equation, or we could write y in terms of x using either equation. Because y has no coefficient in the second equation, we will start by writing y in terms of x using that equation:

$5x + y = 10$
$y = 10 - 5x$

Next, replace y in the first equation with the expression that is equal to y, $10 - 5x$:

$3(10 - 5x) = 12x + 3$

Solve for the value of x:

$3(10 - 5x) = 12x + 3$
$30 - 15x = 12x + 3$
$30 - 15x + 15x = 12x + 15x + 3$
$30 = 27x + 3$
$30 - 3 = 27x + 3 - 3$
$27 = 27x$
$1 = x$

We have the value of one variable, x. Replace x with its value in either equation and solve for the value of y:

$5(1) + y = 10$
$5 + y = 10$
$5 - 5 + y = 10 - 5$
$y = 10 - 5$
$y = 5$

The solution to this system of equations is $x = 1$, $y = 5$.

Practice 1

Solve each system of equations.

1. $x - 2y = -10$, $6x + 2y = -4$

2. $3y - 5x = 6$, $y - x = 4$

3. $8x - 4 = 2y$, $-5x + 3y = -13$

SYSTEMS WITH THREE VARIABLES

In order to find the value of two variables, we need two equations. To find the value of three variables, we need three equations. If we have more than three variables, we need as many equations as we have variables. To find the value of 100 variables, we would need 100 equations!

We solve a system of equations with three variables just as we solve a system of equations with two variables; we just need to use more steps.

$$x + y + z = 3$$
$$3z + 2y = 4x$$
$$5x - z = -1 - y$$

We start by writing one variable in terms of two other variables. We can use the first equation to write any of the variables in terms of the others. Let's write x in terms of y and z:

$$x + y + z = 3$$
$$x = 3 - y - z$$

Next, replace x in each of the other two equations with the expression $3 - y - z$:

$$3z + 2y = 4(3 - y - z)$$
$$3z + 2y = 12 - 4y - 4z$$
$$7z + 2y = 12 - 4y$$
$$7z + 6y = 12$$

$$5(3 - y - z) - z = -1 - y$$
$$15 - 5y - 5z - z = -1 - y$$
$$15 - 4y - 6z = -1$$
$$-4y - 6z = -16$$

We now have two equations and two variables:

$$7z + 6y = 12$$
$$-4y - 6z = -16$$

Write one variable in terms of the other. Let's write y in terms of z using the second equation:

$$-4y - 6z = -16$$
$$-4y = 6z - 16$$
$$y = -\frac{6}{4}z + 4$$

Now, replace y in the first equation with the expression that is equal to y, $-\frac{6}{4}z + 4$:

$$7z + 6(-\frac{6}{4}z + 4) = 12$$
$$7z - 9z + 24 = 12$$
$$-2z + 24 = 12$$
$$-2z = -12$$
$$z = 6$$

We have the value of one variable, z. Replace z with its value in either equation and solve for the value of y:

$$7z + 6y = 12$$
$$7(6) + 6y = 12$$
$$42 + 6y = 12$$
$$6y = -30$$
$$y = -5$$

We have the values of y and z. Substitute them into any one of the original three equations to find the value of x:

$$x + y + z = 3$$
$$x + -5 + 6 = 3$$
$$x + 1 = 3$$
$$x = 2$$

The solution to this system of equations is $x = 2$, $y = -5$, $z = 6$.

Practice 2

Find the solution to this system of equations:

$$3z + 2x = y - 1$$
$$5y - 2z = x + 2$$
$$3x = -4z$$

ANSWERS

Practice 1

1. Start by writing one variable in terms of the other. Because x has no coefficient in the first equation, start by writing x in terms of y using that equation:
$$x - 2y = -10$$
$$x = 2y - 10$$
Replace x in the second equation with the expression that is equal to x, $2y - 10$:
$$6(2y - 10) + 2y = -4$$
Solve for the value of y:
$$6(2y - 10) + 2y = -4$$
$$12y - 60 + 2y = -4$$
$$14y - 60 = -4$$
$$14y = 56$$
$$y = 4$$
Replace y with its value in either equation and solve for the value of x:
$$x - 2(4) = -10$$
$$x - 8 = -10$$
$$x = -2$$
The solution to this system of equations is $x = -2$, $y = 4$.
2. Start by writing one variable in terms of the other. Because y has no coefficient in the second equation, start by writing y in terms of x using that equation:
$$y - x = 4$$
$$y = x + 4$$
Replace y in the first equation with the expression that is equal to y, $x + 4$:
$$3(x + 4) - 5x = 6$$

Solve for the value of x:

$$3(x + 4) - 5x = 6$$
$$3x + 12 - 5x = 6$$
$$-2x + 12 = 6$$
$$-2x = -6$$
$$x = 3$$

Replace x with its value in either equation and solve for the value of y:

$$y - x = 4$$
$$y - 3 = 4$$
$$y = 7$$

The solution to this system of equations is $x = 3$, $y = 7$.

3. Start by writing one variable in terms of the other. Write y in terms of x using the first equation. Divide both sides of the equation by 2:

$$8x - 4 = 2y$$
$$4x - 2 = y$$

Replace y in the second equation with the expression that is equal to y, $4x - 2$:

$$-5x + 3(4x - 2) = -13$$

Solve for the value of x:

$$-5x + 3(4x - 2) = -13$$
$$-5x + 12x - 6 = -13$$
$$7x - 6 = -13$$
$$7x = -7$$
$$x = -1$$

Replace x with its value in either equation and solve for the value of y:

$$8(-1) - 4 = 2y$$
$$-8 - 4 = 2y$$
$$-12 = 2y$$
$$-6 = y$$

The solution to this system of equations is $x = -1$, $y = -6$.

Practice 2

Start by writing one variable in terms of two other variables. Write y in terms of x and z:

$$3z + 2x = y - 1$$
$$3z + 2x + 1 = y$$

Next, replace y in the second equation with the expression $3z + 2x + 1$. The third equation does not have a y at all.

$$5y - 2z = x + 2$$
$$5(3z + 2x + 1) - 2z = x + 2$$
$$15z + 10x + 5 - 2z = x + 2$$
$$13z + 10x + 5 = x + 2$$
$$13z + 9x + 5 = 2$$
$$13z + 9x = -3$$

There are now two equations and two variables:

$$13z + 9x = -3$$
$$3x = -4z$$

Write one variable in terms of the other. Write x in terms of z using the second equation:

$$3x = -4z$$
$$x = \frac{-4}{3}z$$

Replace x in the first equation with the expression that is equal to x, $\frac{-4}{3}z$:

$$13z + 9x = -3$$
$$13z + 9(\frac{-4}{3}z) = -3$$
$$13z - 12z = -3$$
$$z = -3$$

Replace z with its value in either equation and solve for the value of x:

$$3x = -4z$$
$$3x = -4(-3)$$
$$3x = 12$$
$$x = 4$$

Substitute the values of x and z into any one of the original equations to find the value of y:

$$3z + 2x = y - 1$$
$$3(-3) + 2(4) = y - 1$$
$$-9 + 8 = y - 1$$
$$-1 = y - 1$$
$$y = 0$$

The solution to this system of equations is $x = 4$, $y = 0$, $z = -3$.

systems of equations— solving with elimination

To be a scholar of mathematics you must be born with talent, insight, concentration, taste, luck, drive and the ability to visualize and guess.
—PAUL R. HALMOS (1916–2006)
AMERICAN MATHEMATICIAN

In this lesson, you will learn how to solve systems of equations with two or three variables using elimination.

NOW THAT WE KNOW how to solve a system of equations using substitution, let's look at another method: elimination. The goal of elimination is to combine two or more equations together to eliminate a variable, using addition, subtraction, multiplication, or division.

We will start with the same example we looked at in Lesson 18, $3x + y = 7$ and $x - y = 13$. Instead of beginning by writing one variable in terms of the other, we will combine these two equations to eliminate either x or y, leaving us with one equation and one variable.

We can add the equations or subtract one equation from the other. We can also multiply or divide one or both equations by a constant before adding or subtracting. Look closely at these equations:

$$3x + y = 7$$
$$x - y = 13$$

The first equation has a y term and the second equation has a $-y$ term. If we add these two equations together, the y terms will drop out and we will be left with one equation and one variable, x. Add the left side of the first equation to the left side of the second equation and add the right side of the first equation to the right side of the second equation:

$$
\begin{array}{r}
3x + y = 7 \\
+\, x - y = 13 \\
\hline
4x \quad\ = 20
\end{array}
$$

Now we can solve for x by dividing both sides of the equation by 4, and we find that $x = 5$. Once we have the value of one variable, just as with the substitution method, we can use it to find the value of the other variable. Substitute 5 for x in either equation and solve for y:

$$3x + y = 7$$
$$3(5) + y = 7$$
$$15 + y = 7$$
$$y = -8$$

If adding the equations or subtracting one equation from the other will not cause one variable to drop out, we must multiply one equation by a constant and then add or subtract. Look at the following system of equations:

$$2x - 3y = 5$$
$$x + 2y = 13$$

First, we must decide which variable we want to eliminate. Eliminating y will take a couple of steps; before we can add the two equations, the coefficients of y in each equation will have to be the same. We can do that by multiplying the first equation by 2 and by multiplying the second equation by 3. However, we can eliminate x a little more easily. If we multiply just the second equation by 2, we can subtract it from the first equation, and that will eliminate x. We must multiply every term on both sides of the equation by 2:

$$2(x + 2y) = 2(13)$$
$$2x + 4y = 26$$

We do not need to multiply the first equation by anything at all. It is important to understand that what we do to one equation we *do not* have to do to the other equation. If we choose to multiply an equation by a constant, we must multiply every term in the equation by the constant. Because every term grows in exactly the same way, the value of the equation does not change. The equation $2x + 4y = 26$ is still true after being multiplied by 2, and the equation $2x - 3y = 5$ is still true after not being changed at all.

Now, we can subtract the second equation from the first:

$$
\begin{array}{r}
2x - 3y = 5 \\
- (2x + 4y = 26) \\
\hline
-7y = -21
\end{array}
$$

We can divide both sides of the equation by –7, and we find that $y = 3$. Substitute 3 for y in either equation and solve for x:

$$x + 2y = 13$$
$$x + 2(3) = 13$$
$$x + 6 = 13$$
$$x = 7$$

The solution to this system of equations is $x = 7$, $y = 3$.

....

TIP: Now that you know two methods for solving systems of equations, you can use either one to solve any system. If an equation in a system has a variable alone on one side of the equation, such as $y = 2x + 4$, substitution might be the easier route to take. If the two equations have an identical variable term, such as $-3x$, elimination might be the easier method. Either method will work every time!

....

Practice 1

Solve each system of equations.

1. $8y + 3x = -2$, $-8y - 14 = 5x$

2. $x - 2y = 20$, $4x + 7y = 5$

3. $-4x + 3y = -4$, $6x - 5y = 2$

SYSTEMS WITH THREE VARIABLES

We can also solve systems of equations with three variables using the elimination method. Just as with the substitution method, we begin by reducing three equations with three variables to two equations with two variables. In fact, with the elimination method, sometimes we can eliminate two variables at once.

$$x - y + 2z = 1$$
$$3x + 1 = 7y - 2z$$
$$-x + 2y - 2z = 3$$

The first equation has an x term and the third equation has an $-x$ term. Adding these equations will eliminate the variable x. But look again—the first equation has a $2z$ term and the third equation has a $-2z$ term. When we add, we will eliminate the variable z, too:

$$
\begin{array}{r}
x - \ y + 2z = 1 \\
+ \ -x + 2y - 2z = 3 \\
\hline
y \quad\quad = 4
\end{array}
$$

We have the value of y, but we still need to find the values of the other two variables. Replace y with 4 in each equation:

$x - y + 2z = 1$	$3x + 1 = 7y - 2z$	$-x + 2y - 2z = 3$
$x - 4 + 2z = 1$	$3x + 1 = 7(4) - 2z$	$-x + 2(4) - 2z = 3$
$x + 2z = 5$	$3x + 1 = 28 - 2z$	$-x + 8 - 2z = 3$
	$3x + 2z = 27$	$-x - 2z = -5$
		$x + 2z = 5$

We have three equations, but the first and third equations are the same. Combine the first two equations. Both have a $2z$ term, so we can subtract the second equation from the first:

$$
\begin{array}{r}
x + 2z = \ \ 5 \\
- \ 3x + 2z = \ 27 \\
\hline
-2x \quad\quad = -22
\end{array}
$$

Divide both sides of the equation by –2, and we find that $x = 11$. Now that we have the values of x and y, we can substitute them into any of the original three equations to find the value of z:

$$x - y + 2z = 1$$
$$(11) - (4) + 2z = 1$$
$$7 + 2z = 1$$
$$2z = -6$$
$$z = -3$$

The solution to this system of equations is $x = 11$, $y = 4$, $z = -3$.

Practice 2

Find the solution to this system of equations:

$$-x + y + 3z = 1$$
$$2y + 9x + 4z = -1$$
$$-8z - 4x = 3y$$

ANSWERS

Practice 1

1. Choose a variable to eliminate. The first equation contains an $8y$ term and the second equation contains a $-8y$ term. Add these equations to eliminate y:

$$
\begin{array}{r}
8y + 3x \quad\quad = \quad -2 \\
+\,-8y \quad\quad - 14 = 5x \\
\hline
3x - 14 = 5x - 2
\end{array}
$$

Solve for x:

$$3x - 14 = 5x - 2$$
$$-14 = 2x - 2$$
$$-12 = 2x$$
$$-6 = x$$

Substitute –6 for x in either equation and solve for y:

$$8y + 3x = -2$$
$$8y + 3(-6) = -2$$
$$8y - 18 = -2$$
$$8y = 16$$
$$y = 2$$

The solution to this system of equations is $x = -6$, $y = 2$.

2. Choose a variable to eliminate. Adding the two equations will not eliminate a variable, and subtracting one equation from the other will not eliminate a variable, either. To eliminate y, we would have to multiply both equations by constants before adding. To eliminate x, we can multiply just the first equation by –4 before adding:

$$-4(x - 2y) = -4(20)$$
$$-4x + 8y = -80$$

The first equation now contains a $-4x$ term and the second equation contains a $4x$ term. Add these equations to eliminate x:

$$-4x + 8y = -80$$
$$\underline{+\ 4x + 7y = \quad 5}$$
$$15y = -75$$

Divide by 15 to solve for y:

$$15y = -75$$
$$y = -5$$

Substitute –5 for y in either equation and solve for x:

$$x - 2y = 20$$
$$x - 2(-5) = 20$$
$$x + 10 = 20$$
$$x = 10$$

The solution to this system of equations is $x = 10$, $y = -5$.

3. Choose a variable to eliminate. Adding the two equations will not eliminate a variable, and subtracting one equation from the other will not eliminate a variable, either. To eliminate x, we can multiply the first equation by 3 and multiply the second equation by 2 before adding:

$$3(-4x + 3y) = 3(-4) \qquad\qquad 2(6x - 5y) = 2(2)$$
$$-12x + 9y = -12 \qquad\qquad\quad 12x - 10y = 4$$

The first equation now contains a $-12x$ term and the second equation contains a $12x$ term. Add these equations to eliminate x:

$$-12x + 9y = -12$$
$$\underline{+\ 12x - 10y = \quad 4}$$
$$-y = -8$$

Divide by –1 to solve for y:

$-y = -8$

$y = 8$

Substitute 8 for y in either equation and solve for x:

$-4x + 3y = -4$

$-4x + 3(8) = -4$

$-4x + 24 = -4$

$-4x = -28$

$x = 7$

The solution to this system of equations is $x = 7$, $y = 8$.

Practice 2

Choose a variable to eliminate. Adding any two of the three equations will not eliminate a variable, and subtracting one equation from another will not eliminate a variable, either. To eliminate y, we can multiply the first equation by –2 and add it to the second equation:

$-2(-x + y + 3z) = -2(1)$

$2x - 2y - 6z = -2$

$$\begin{array}{r} 2x - 2y - 6z = -2 \\ + 9x + 2y + 4z = -1 \\ \hline 11x \quad\quad - 2z = -3 \end{array}$$

We can combine another pair of equations so that we have a second equation with just the variables x and z. Multiply the second equation by 3 and third equation by –2:

$3(2y + 9x + 4z) = 3(-1)$ $-2(-8z - 4x) = -2(3y)$

$6y + 27x + 12z = -3$ $16z + 8x = -6y$

Add the equations to eliminate the variable y:

$$\begin{array}{r} 6y + 27x + 12z = \quad\quad\quad -3 \\ + -6y \quad\quad\quad\quad = 16z + 8x \\ \hline 27x + 12z = 16z + 8x - 3 \end{array}$$

Simplify:

$$27x + 12z = 16z + 8x - 3$$
$$19x + 12z = 16z - 3$$
$$19x - 4z = -3$$

We have two equations with x and z:

$$11x - 2z = -3$$
$$19x - 4z = -3$$

Multiply the first equation by 2 and then subtract the second equation:

$$2(11x - 2z) = 2(-3)$$
$$22x - 4z = -6$$

$$22x - 4z = -6$$
$$\underline{-\ 19x - 4z = -3}$$
$$3x\qquad = -3$$
$$x = -1$$

We have the value of x. Substitute it into one of the two equations with just x and z:

$$11x - 2z = -3$$
$$11(-1) - 2z = -3$$
$$-11 - 2z = -3$$
$$-2z = 8$$
$$z = -4$$

Substitute the values of x and z into any one of the original equations to find the value of y:

$$-x + y + 3z = 1$$
$$-(-1) + y + 3(-4) = 1$$
$$1 + y - 12 = 1$$
$$y - 11 = 1$$
$$y = 12$$

The solution to this system of equations is $x = -1$, $y = 12$, $z = -4$.

algebraic inequalities

Mathematics is like checkers in being suitable for the young, not too difficult, amusing, and without peril to the state.

—PLATO (C. 424/423–348/347 B.C.)
CLASSICAL GREEK PHILOSOPHER

In this lesson, you'll learn how to solve single-variable and compound inequalities, and how to simplify algebraic inequalities with two variables.

IF AN EQUATION is what we write to show two quantities that are equal to each other, what can we write to show that two quantities are NOT equal to each other? An inequality. Inequalities can use the less than sign (<), the greater than sign (>), the less than or equal to sign (≤), and the greater than or equal to sign (≥) to compare two quantities. An **algebraic inequality** is an algebraic expression that contains one of those four signs.

We solve single-variable equations by isolating the variable on one side of the equal sign and its value on the other side. We solve single-variable inequalities in the same way, except that instead of finding an answer that is a single value, our answer is a set of values.

The equation $x + 4 = 9$ is solved by subtracting 4 from both sides of the equal sign: $x + 4 - 4 = 9 - 4$, $x = 5$. The inequality $x + 4 < 9$ is solved in the same way: Subtract 4 from both sides of the less than sign: $x + 4 - 4 < 9 - 4$, $x < 5$. Our answer is $x < 5$, which means that all values of x that are less than 5 make the inequality true.

We can represent an inequality on a number line. To show $x < 5$, we put a circle around the number 5, because it is not a part of our answer (our solution is only values of x that are less than 5), and we highlight all of the values to the left of 5, to show that every number that is less than 5 is part of the solution:

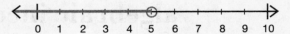

There is one important difference between how we solve an equation and how we solve an inequality. When you are solving an inequality, if you multiply or divide both sides of the equation by a negative number, you must flip the inequality symbol. For example, to solve $-5x < 25$, divide both sides of the inequality by -5. When you do that, switch the inequality symbol from the less than sign to the greater than sign:

$-5x < 25$

$x > 5$

Why do we switch the symbol? Let's look at some real numbers. We know that -1 is less than 2, and we show that by writing $-1 < 2$. If we multiply both sides of the inequality by 2, we have $-2 < 4$, which is also true. The left side of the inequality became twice as small, and the right side of the inequality became twice as large. But what if we were to divide both sides of $-1 < 2$ by -1. The left side would become 1, and the right side would become -2. However, 1 is greater than -2. So, we must switch the less than sign to a greater than sign, to show that $1 > -2$.

Let's look at another example: $3x - 7 \geq 2$. Add 7 to both sides and divide by 3:

$3x - 7 \geq 2$

$3x \geq 9$

$x \geq 3$

The number line of this inequality shows a solid circle around 3, because 3 is part of the solution set (since 3 is greater than or equal to 3):

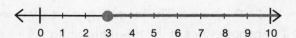

TIP: We cannot divide both sides of an inequality by a variable, because we do not know if the variable is positive or negative. The inequalities $x^2 > 2x$ and $x > 2$ are not the same, because $x = -3$ would be in the solution set of the first inequality, but not the second inequality.

Practice 1

Solve each inequality for x. Show the solution on a number line with values from −10 to 10.

1. $6x + 11 < 29$

2. $\frac{1}{4}x \geq -2$

3. $8 - 7x > 50$

4. $3x - 6 < 0$

5. $-4x + 7 \leq 19$

COMPOUND INEQUALITIES

So far, we have seen inequalities where a variable is less than, less than or equal to, greater than, or greater than or equal to a quantity. Sometimes, a variable is between two quantities. For example, if x can be as small as −4 but less than 7, we would write $-4 \leq x < 7$. The number −4 is part of the solution because x can be −4, but 7 is not part of the solution because x is less than 7. We can show this inequality on a number line, too. No arrows are highlighted, because the solution set has two boundaries:

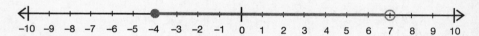

An equation and a simple inequality both have left and right sides. Whatever operation we perform on one side, we perform on the other. A compound inequality has three parts, since it has two inequality symbols. Whatever operation we perform on one part, we must perform on all three parts.

To simplify the compound inequality $7 < x + 2 < 9$, isolate the variable, x, in the center of the inequality. Do this by subtracting 2 from all three parts of the inequality, which gives us $5 < x < 7$.

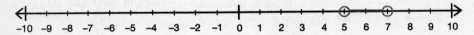

To simplify the inequality $-4 \leq -2x \leq 10$, we must divide each part of the inequality by -2. Because we are dividing by a negative number, we must change both inequality symbols: $-4 \leq -2x \leq 10$ becomes $2 \geq x \geq -5$.

Practice 2

Solve each compound inequality for x. Show the solution on a number line with values from -10 to 10.

1. $0 < x - 4 \leq 4$

2. $49 > -7x > -21$

3. $-13 \leq 3x - 7 < 20$

4. $1 \geq \frac{1}{5}x + 1 \geq -1$

5. $-1 > -\frac{1}{2}x - 5 \geq -3$

INEQUALITIES WITH TWO VARIABLES

When we are given an inequality with both x and y, we simplify it by writing y in terms of x.

Given the inequality $12x + 3y < 6$, we subtract $12x$ from both sides and divide by 3:

$$12x + 3y < 6$$
$$3y < 6 - 12x$$
$$y < 2 - 4x$$

Just as with single-variable inequalities, if we multiply or divide both sides of the inequality by a negative number, we must change the inequality symbol. To solve $x - 5y \geq 15$ for y in terms of x, we subtract x from both sides, divide by -5, and change the inequality symbol from $\geq$ to $\leq$:

$$x - 5y \geq 15$$
$$-5y \geq 15 - x$$
$$y \leq -3 + \tfrac{1}{5}x$$

Practice 3

Solve each inequality for y in terms of x.

1. $y - 4 > x$

2. $8y - 2x \leq 24$

3. $15x - 5y > 100$

ANSWERS

Practice 1

1. Subtract 11 from both sides of the inequality and divide by 6:
$$6x + 11 < 29$$
$$6x < 18$$
$$x < 3$$
Because x is less than 3, 3 is not part of the solution set. Draw a hollow circle around 3 and shade to the left, where values are less than 3:

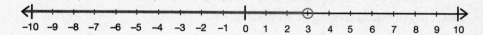

2. Multiply both sides of the inequality by 4:
$$\tfrac{1}{4}x \geq -2$$
$$4\left(\tfrac{1}{4}x\right) \geq 4(-2)$$
$$x \geq -8$$

Because x is greater than or equal to –8, –8 is part of the solution set. Draw a closed circle around –8 and shade to the right, where values are greater than –8:

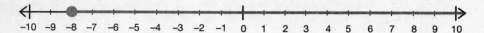

3. Subtract 8 from both sides of the inequality and divide by –7. Because we are dividing by a negative number, change the inequality symbol from greater than to less than:

$$8 - 7x > 50$$
$$-7x > 42$$
$$x < -6$$

Because x is less than –6, –6 is not part of the solution set. Draw a hollow circle around –6 and shade to the left, where values are less than –6:

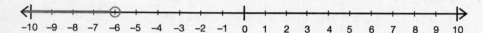

4. Add 6 to both sides of the inequality and divide by 3:

$$3x - 6 < 0$$
$$3x < 6$$
$$x < 2$$

Because x is less than 2, 2 is not part of the solution set. Draw a hollow circle around 2 and shade to the left, where values are less than 2:

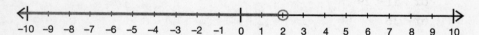

5. Subtract 7 from both sides of the inequality and divide by –4. Because we are dividing by a negative number, change the inequality symbol from less than or equal to greater than or equal:

$$-4x + 7 \leq 19$$
$$-4x \leq 12$$
$$x \geq -3$$

Because x is greater than or equal to –3, –3 is part of the solution set. Draw a closed circle around –3 and shade to the right, where values are greater than –3:

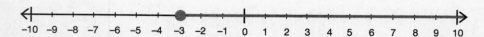

Practice 2

1. Add 4 to each part of the inequality:
$$0 < x - 4 \leq 4$$
$$4 < x \leq 8$$
Because x is greater than 4, 4 is not part of the solution set, and because x is less than or equal to 8, 8 is part of the solution set. Draw a hollow circle around 4 and a closed circle around 8. Shade the area between them:

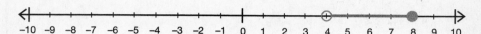

2. Divide by –7 and change the inequality symbols from greater than to less than:
$$49 > -7x > -21$$
$$-7 < x < 3$$
Because x is greater than –7, –7 is not part of the solution set, and because x is less 3, 3 is not part of the solution set. Draw a hollow circle around –7 and a hollow circle around 3. Shade the area between them:

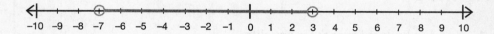

3. Add 7 to each part of the inequality and divide by 3:
$$-13 \leq 3x - 7 < 20$$
$$-6 \leq 3x < 27$$
$$-2 \leq x < 9$$
Because x is greater than or equal to –2, –2 is part of the solution set, and because x is less than 9, 9 is not part of the solution set. Draw a closed circle around –2 and a hollow circle around 9. Shade the area between them:

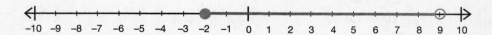

4. Subtract 1 from each part of the inequality and multiply by 5:
$$1 \geq \tfrac{1}{5}x + 1 \geq -1$$
$$0 \geq \tfrac{1}{5}x \geq -2$$
$$0 \geq x \geq -10$$
Because x is less than or equal to 0, 0 is part of the solution set, and because x is greater than or equal to –10, –10 is part of the solution set.

Draw a closed circle around 0 and a closed circle around –10. Shade the area between them:

5. Add 5 to each part of the inequality and multiply by –2. Change the inequality symbols from greater than and greater than or equal to to less than and less than or equal:

$$-1 > -\tfrac{1}{2}x - 5 \geq -3$$
$$4 > -\tfrac{1}{2}x \geq 2$$
$$-8 < x \leq -4$$

Because x is greater than –8, –8 is not part of the solution set, and because x is less than or equal to –4, –4 is part of the solution set. Draw a hollow circle around –8 and a closed circle around –4. Shade the area between them:

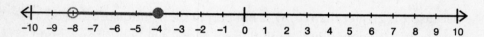

Practice 3

1. Add 4 to both sides of the inequality:
$$y - 4 > x$$
$$y > x + 4$$

2. Add $2x$ to both sides of the inequality and divide by 8:
$$8y - 2x \leq 24$$
$$8y \leq 2x + 24$$
$$y \leq \tfrac{1}{4}x + 3$$

3. Subtract $15x$ from both sides of the equation. Divide by –5 and change the inequality symbol from greater than to less than:
$$15x - 5y > 100$$
$$-5y > 100 - 15x$$
$$y < -20 + 3x$$

graphing inequalities

A line is a dot that went for a walk.
—PAUL KLEE (1879–1940)
SWISS PAINTER

In this lesson, you'll learn how to graph inequalities with two variables and how to graph the solution set of a system of inequalities.

THE SOLUTIONS OF THE EQUATION of a line are each point on the graph of the line. For the line $y = x - 1$, the point (2,1) is on the line, and the point (2,1) is a solution of the equation. An inequality can have many more solutions than the equation of a line. The inequality $y < x + 5$ is true whenever y is less than five more than x. When x is 3, y can be any number that is less than 8. When x is 4, y can be any number that is less than 9. To show the solution of an inequality, we graph a line and then shade the side of the line where all of the solutions lie.

In order to graph $y < x + 5$, we must first graph the line $y = x + 5$. The point (1,6) is on that line, but is (1,6) a solution of $y < x + 5$? It is not, because 6 is not less than $1 + 5$. The points on the line $y = x + 5$ are not part of the solution set of $y < x + 5$, so we graph a dashed line instead of a solid line.

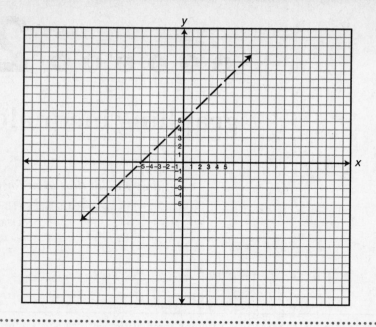

....

TIP: When an inequality contains the < or > symbol, the graph of the line will be dashed. When an inequality contains the ≤ or ≥ symbol, the graph of the line will be solid.

....

Which side of the line contains the solutions to $y < x + 5$? We pick a test point to decide which side of the line to shade. The point (0,0) is the best point to choose, because it makes our calculations easy. Substitute 0 for y and 0 for x:

$$y < x + 5$$
$$0 < 0 + 5 ?$$
$$0 < 5$$

Because 0 is less than 5, the point (0,0) is part of the solution set. Shade the area below the line to show that this is where the solutions lie.

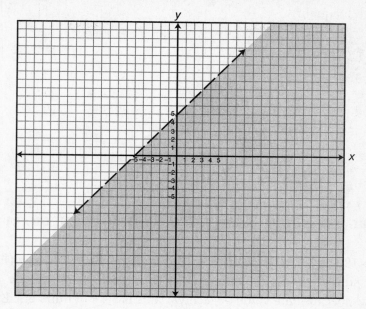

We can also graph inequalities with one variable. To graph $y \geq -4$, we start by graphing the line $y = -4$. Every value on the line has a y value of -4, and because $-4 \leq -4$, the points on this line are part of the solution set. We will make the graph of the line solid. Next, test the point (0,0) to see if it is part of the solution set. Because 0 is greater than or equal to -4, (0,0) is in the solution set. Shade the area above the line.

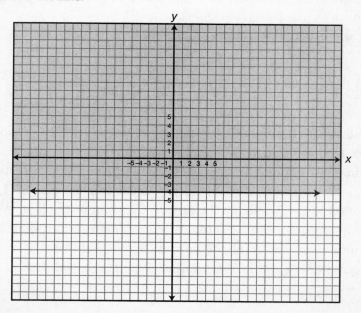

Practice 1

Graph each inequality.

1. $y \leq 2x + 7$

2. $y < -4x - 3$

3. $x > 8$

4. $x - y \leq -2$

GRAPHING THE SOLUTION SET OF A SYSTEM OF INEQUALITIES

In Lessons 18 and 19, we learned two methods for solving a system of equations. However, substitution and elimination will not work for solving a system of inequalities, because the value of each variable is a set of numbers. Even if we could isolate a variable by itself, we could never substitute a single value for it in order to find the value of the other variables. We must use a graph to show the solution of a system of inequalities.

If we are given two inequalities, we can plot them both and shade the solution of each on the same set of axes. Our solution to the system is the overlapping area. If the two inequalities have no overlapping area, then the system has no solution.

Look at the following system of inequalities.

$$y < 5x$$
$$y > -x - 1$$

To find the solution to this system, we begin by graphing $y = 5x$ and $y = -x - 1$. Both lines will be dashed since the inequality symbols are less than and greater than. The test point $(0,0)$ cannot be used on the first inequality, because that point is on the line $y = 5x$. Try $(1,1)$: Because 1 is less than 5, $(1,1)$ is part of the solution to $y < 5x$. Shade the area to the right of the line. We can test $y > -x - 1$ with the point $(0,0)$. Because 0 is greater than -1, the point $(0,0)$ is part of the solution to $y > -x - 1$. Shade the area above that line. The overlapping area is in quadrants I and IV. This darker region is the solution to the set of inequalities.

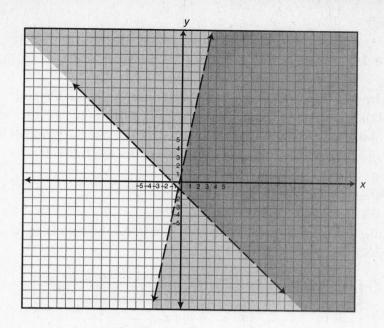

TIP: Sometimes, the only difference between two inequalities in a system is the inequality symbol, as in the system $y < x + 1$ and $y > x + 1$. This system has no solution, because there are no y values that can be less than $x + 1$ and greater than $x + 1$. If the two inequalities were $y \leq x + 1$ and $y \geq x + 1$, the solution would be all the values that are on the line $y = x + 1$, since that is the only area the two inequalities have in common. To sum up, when the only difference between two inequalities is the inequality symbol, if the symbols are $\leq$ and $\geq$, then the solution is all points on the graphed line. If the symbols are $<$ and $>$, then the system has no solution.

Practice 2

Graph the solution to each system of inequalities.

1. $y > x, x < -3$

2. $y \geq -\frac{1}{2}x + 2, y < -3x - 3$

ANSWERS

Practice 1

1. Start by graphing the line $y = 2x + 7$. The point (1,9) is on the graph of that line. It is true that $9 \leq 2(1) + 7$, because $9 \leq 9$, so the points on this line are part of the solution to the inequality $y \leq 2x + 7$. Make the graph of the line solid. Test the point (0,0) to see if it is part of the solution set:

$$0 \leq 2(0) + 7 ?$$
$$0 \leq 0 + 7 ?$$
$$0 \leq 7$$

Because 0 is less than or equal to 7, the point (0,0) is part of the solution set. Shade the area to the right of the line.

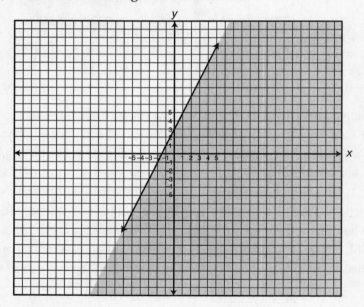

2. Start by graphing the line $y = -4x - 3$. The point (0,–3) is on the graph of that line. It is not true that $-3 < -4(0) - 3$, because –3 is not less than –3, so the points on this line are not part of the solution to the inequality $y < -4x - 3$. Make the graph of the line dashed. Test the point (0,0) to see if it is part of the solution set:

$$0 < -4(0) - 3?$$
$$0 < 0 - 3 ?$$
$$0 < -3 ?$$

Because 0 is not less than –3, the point (0,0) is not part of the solution set. Shade the area to the left of the line.

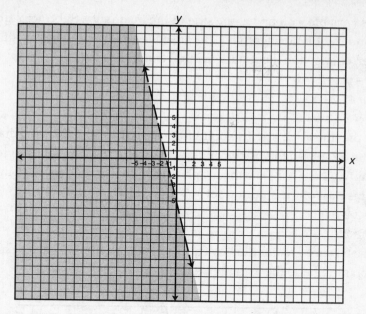

3. Start by graphing the line $x = 8$. Every point on that line has an x value of 8. Because 8 is not greater than 8, the points on this line are not part of the solution to the inequality $x > 8$. Make the graph of the line dashed. Test the point $(0,0)$ to see if it is part of the solution set:

$$0 > 8 ?$$

Because 0 is not greater than 8, the point $(0,0)$ is not part of the solution set. Shade the area to the right of the line.

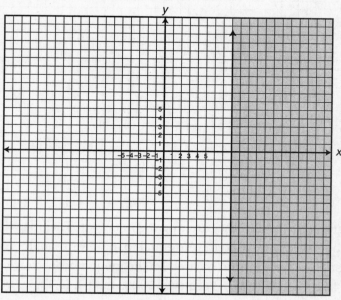

4. Get the variable y alone on one side of the inequality. Subtract x from both sides and divide by negative 1. Because we are dividing by a negative number, change the inequality symbol from $\le$ to $\ge$:

$$x - y \le -2$$
$$-y \le -2 - x$$
$$y \le 2 + x$$

Graph the line $y = 2 + x$. The point (0,2) is on that line. It is true that $2 \ge 2 + 0$, because $2 \ge 2$, so the points on this line are part of the solution to the inequality $y \ge 2 + x$. Make the graph of the line solid. Test the point (0,0) to see if it is part of the solution set:

$$0 \ge 2 + 0 ?$$
$$0 \ge 2 ?$$

Because 0 is not greater than or equal to 2, the point (0,0) is not part of the solution set. Shade the area above the line.

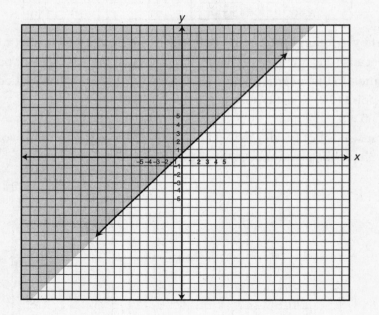

Practice 2

1. First, graph $y > x$. Start by graphing the line $y = x$. The line will be dashed since the x values on that line are equal to, not less than, their y values. The test point (0,0) cannot be used on the inequality, because that point is on the line $y = x$. Try (0,4): Because 0 is less than 4, (0,4) is part of the solution to $y > x$. Shade the area above the line.

Next, graph $x < -3$. Start by graphing the line $x = -3$. The line will be dashed since the x values on that line are equal to -3, not less than -3. Use (0,0) as a test point to decide which side of the line to shade. Because 0 is not less than -3, (0,0) is not part of the solution to $x < -3$. Shade the area to the left of the line.

The overlapping area is in quadrants II and III. The darker region is the solution to the system.

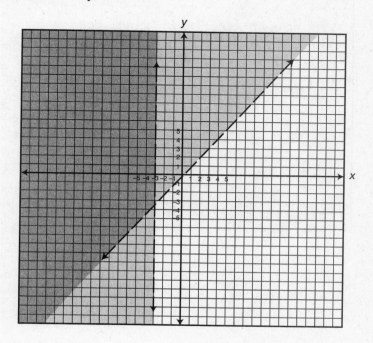

2. First, graph $y \geq -\frac{1}{2}x + 2$. Start by graphing the line $y = -\frac{1}{2}x + 2$. The line will be solid because the inequality contains the greater than or equal to symbol. Use (0,0) as a test point to decide which side of the line to shade. Because 0 is not greater than or equal to 2, (0,0) is not part of the solution to $y \geq -\frac{1}{2}x + 2$. Shade the area above the line.

Next, graph $y < -3x - 3$. Start by graphing the line $y = -3x - 3$. The line will be dashed because the inequality contains the less than symbol. Use (0,0) as a test point. Because 0 is not less than -3, (0,0) is not part of the solution to $y < -3x - 3$. Shade the area to the left of the line.

The overlapping area is in quadrant II. The darker region is the solution to the system.

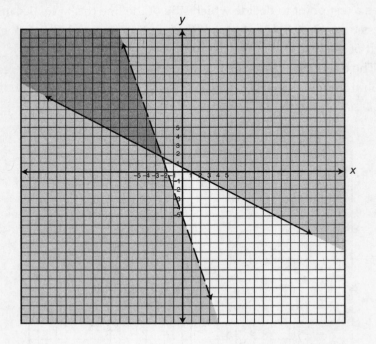

polynomials and FOIL

I'm kinda tired. I was up all night trying to round off infinity.
—STEVEN WRIGHT (1955–   )
AMERICAN COMEDIAN

In this lesson, you'll learn how to describe algebraic polynomials and how to multiply binomials.

BY NOW, WE'RE PRETTY COMFORTABLE MULTIPLYING one term by another. An expression that is one term or the sum of two or more terms, each with whole numbered exponents, is called a **polynomial**. A polynomial that contains only one term, such as $3x^2$, is called a **monomial**. A polynomial that is the sum of two terms, such as $2x + 1$, is called a **binomial**, and a polynomial that is the sum of three terms, such as $x^2 + 4x + 4$, is called a trinomial.

The expression $5x^2 + 2x + 1$ is a polynomial. It contains an x^2 term, an x term, and a constant. Polynomials have exponents that are whole numbers—never fractions, variables, or negative numbers. The coefficients of polynomials can be fractions or negative numbers. The sum of two or more polynomials is a polynomial, and the product of two or more polynomials is a polynomial.

We can classify a polynomial by the degrees of its variables. The **degree** of a variable is its exponent. Constants have a degree of zero. If a polynomial consists of only a constant, then the polynomial has a degree of zero and is classified as a constant. A polynomial such as $x + 2$ contains a variable with an exponent of one and a constant. This polynomial has a degree of one and is classified as

linear. The polynomial $x^2 - 5$ has a degree of two, and it is classified as **quadratic**. Polynomials such as $x^3 + 1$ have a degree of 3, and they are classified as **cubic**. In these next few lessons, we'll be learning about quadratic equations.

MULTIPLYING BINOMIALS

We've seen how to multiply two monomials, such as $2x$ and $4x^2$, by multiplying their coefficients and adding the exponents of common bases:

$$(2x)(4x^2) = 8x^3$$

We've also seen how to multiply a monomial by a binomial using the distributive law. Each term in the binomial is multiplied by the monomial:

$$4x(5x + 4) = 20x^2 + 16x$$

To multiply two binomials, we must multiply each term in the first binomial by each term in the second binomial. This will give us four products, which we will combine if there are any like terms.

Example
$(2x + 1)(3x + 4)$

We use the acronym **FOIL** to help us remember how to multiply binomials. FOIL stands for *First Outside Inside Last*. We begin by multiplying the First terms in each binomial. In this example, the first term in the first binomial is $2x$ and the first term in the second binomial is $3x$:

$$(2x)(3x) = 6x^2$$

Next, we multiply the Outside terms. The outside terms are the first term in the first binomial and the last term in the second binomial, because when the binomials are written as $(2x + 1)(3x + 4)$, these are the terms on the outside of the expression. In this example, the outside terms are $2x$ and 4:

$$(2x)(4) = 8x$$

Then, we multiply the Inside terms. The inside terms are the second term in the first binomial and first term in the second binomial. When the binomials are written as $(2x + 1)(3x + 4)$, these terms appear next to each other on the inside of the expression. In this example, the inside terms are 1 and $3x$:

$(1)(3x) = 3x$

Finally, we multiply the Last terms in each binomial. In this example, the last terms are 1 and 4:

$(1)(4) = 4$

We have finished multiplying, and we have four products to sum and combine any like terms: $6x^2 + 8x + 3x + 4$. Two of these terms, $8x$ and $3x$, are like terms, so they can be combined: $8x + 3x = 11x$. The product of $(2x + 1)$ and $(3x + 4)$ is $6x^2 + 11x + 4$.

....

TIP: When two linear binomials are multiplied, we will often be left with one product that has a degree of two, two products with a degree of one, and a constant. The two products with a degree of one can often be combined, leaving us with a trinomial for an answer.

....

Example

$(5x + 5)(6x - 4)$

First: $(5x)(6x) = 30x^2$
Outside: $(5x)(-4) = -20x$
Inside: $(5)(6x) = 30x$
Last: $(5)(-4) = -20$

Combine like terms: $-20x + 30x = 10x$, so $(5x + 5)(6x - 4) = 30x^2 + 10x - 20$.

SQUARING A BINOMIAL

We can use FOIL to find the square of a binomial, too. Remember, an exponent just states how many times a quantity must be multiplied. When you see an expression such as $(x + 1)^2$, rewrite it as $(x + 1)(x + 1)$. Now use FOIL:

First: $(x)(x) = x^2$
Outside: $(x)(1) = x$

Inside: $(1)(x) = x$

Last: $(1)(1) = 1$

Combine like terms: $x + x = 2x$, so $(x + 1)^2 = (x + 1)(x + 1) = x^2 + 2x + 1$.

So far, the product of two binomials has always been a trinomial. But if the products of the inside and outside terms are **additive inverses**, such as $4x$ and $-4x$, our answer will be a binomial—just a term of degree two and a constant. A quantity and its negative are additive inverses.

$(x - 2)(4x + 8)$

First: $(x)(4x) = 4x^2$

Outside: $(x)(8) = 8x$

Inside: $(-2)(4x) = -8x$

Last: $(-2)(8) = -16$

When like terms are combined, $8x - 8x = 0$. The product has no x term: $4x^2 - 16$.

Practice

Find the product of each binomial.

1. $(x + 2)(x + 3)$

2. $(2x - 1)(x - 5)$

3. $(3x + 7)(6x - 10)$

4. $(-x + 3)(8x + 15)$

5. $(7x + 2)(2x + 4)$

6. $(x + 9)(9x - 8)$

7. $(x - 8)^2$

8. $(3x + 2)^2$

9. $(x + 10)(x - 10)$

10. $(5x - 12)(5x + 12)$

ANSWERS

Practice

1. Use FOIL and combine like terms:
   First: $(x)(x) = x^2$
   Outside: $(x)(3) = 3x$
   Inside: $(2)(x) = 2x$
   Last: $(2)(3) = 6$
   $x^2 + 3x + 2x + 6 = x^2 + 5x + 6$

2. Use FOIL and combine like terms:
   First: $(2x)(x) = 2x^2$
   Outside: $(2x)(-5) = -10x$
   Inside: $(-1)(x) = -x$
   Last: $(-1)(-5) = 5$
   $2x^2 - 10x - x + 5 = 2x^2 - 11x + 5$

3. Use FOIL and combine like terms:
   First: $(3x)(6x) = 18x^2$
   Outside: $(3x)(-10) = -30x$
   Inside: $(7)(6x) = 42x$
   Last: $(7)(-10) = -70$
   $18x^2 - 30x + 42x - 70 = 18x^2 + 12x - 70$

4. Use FOIL and combine like terms:
   First: $(-x)(8x) = -8x^2$
   Outside: $(-x)(15) = -15x$
   Inside: $(3)(8x) = 24x$
   Last: $(3)(15) = 45$
   $-8x^2 - 15x + 24x + 45 = -8x^2 + 9x + 45$

5. Use FOIL and combine like terms:
   First: $(7x)(2x) = 14x^2$
   Outside: $(7x)(4) = 28x$
   Inside: $(2)(2x) = 4x$
   Last: $(2)(4) = 8$
   $14x^2 + 28x + 4x + 8 = 14x^2 + 32x + 8$

6. Use FOIL and combine like terms:
   First: $(x)(9x) = 9x^2$
   Outside: $(x)(-8) = -8x$
   Inside: $(9)(9x) = 81x$
   Last: $(9)(-8) = -72$
   $9x^2 - 8x + 81x - 72 = 9x^2 + 73x - 72$

7. Because $(x - 8)$ is raised to the second power, multiply it by itself: $(x - 8)^2 =$ $(x - 8)(x - 8)$. Use FOIL and combine like terms:

First: $(x)(x) = x^2$

Outside: $(x)(-8) = -8x$

Inside: $(-8)(x) = -8x$

Last: $(-8)(-8) = 64$

$x^2 - 8x - 8x + 64 = x^2 - 16x + 64$

8. Because $(3x + 2)$ is raised to the second power, multiply it by itself: $(3x + 2)^2 = (3x + 2)(3x + 2)$. Use FOIL and combine like terms:

First: $(3x)(3x) = 9x^2$

Outside: $(3x)(2) = 6x$

Inside: $(2)(3x) = 6x$

Last: $(2)(2) = 4$

$9x^2 + 6x + 6x + 4 = 9x^2 + 12x + 4$

9. Use FOIL and combine like terms:

First: $(x)(x) = x^2$

Outside: $(x)(-10) = -10x$

Inside: $(10)(x) = 10x$

Last: $(10)(-10) = -100$

$x^2 - 10x + 10x - 100 = x^2 - 100$

10. Use FOIL and combine like terms:

First: $(5x)(5x) = 25x^2$

Outside: $(5x)(12) = 60x$

Inside: $(-12)(5x) = -60x$

Last: $(-12)(12) = -144$

$25x^2 + 60x - 60x - 144 = 25x^2 - 144$

quadratic factoring

Life is good for only two things, discovering
mathematics and teaching mathematics.
—Siméon Poisson (1781–1840)
French mathematician

In this lesson, you'll learn how to factor a trinomial into two binomials, and you'll learn how to factor the difference of perfect squares into two binomials.

LESSON 7 TAUGHT US HOW TO FACTOR A MONOMIAL out of a polynomial. We found the greatest common factor of the coefficients of each term, and we found the variables common to every term in the polynomial. Now, we will learn how to factor quadratic expressions into two binomials. Remember, a quadratic expression is an expression where the highest degree of a variable is 2.

We use FOIL to multiply two binomials. The result is often a trinomial. To break that trinomial back down into two binomials, we must do the reverse of FOIL. The expression $x^2 + 5x + 6$ has three terms: an x^2 term, an x term, and a constant. We start by setting up parentheses to show the two binomials that will be our factors:

$$(\_ + \_)(\_ + \_)$$

We must find the two terms that make up the first binomial and the two terms that make up the second binomial. Of the three terms in $x^2 + 5x + 6$, it is the middle term, $5x$, that is the trickiest, because it is likely the sum of two terms. Start by listing the factors of the first term and the last term.

Factors of x^2: x

Factors of 6: 1, 2, 3, 6

The only way to form x^2 is to multiply x by itself. Make the first term in each binomial x:

$(x + \_\_)(x + \_\_)$

The number 6 can be found by multiplying 1 and 6 or 2 and 3. When we list the factors of a number, we usually list only the positive factors, but when factoring a quadratic expression, it is just as likely that two negative numbers are the factors. The complete list of integer factors of 6 is –6, –3, –2, –1, 1, 2, 3, and 6.

We can use the middle term, $5x$, to help us decide which factors to choose for 6. If the factors of 6 were negative, then when FOIL was used to make $x^2 + 5x + 6$, the Outside and Inside products would be negative, and the middle term of $x^2 + 5x + 6$ would be negative. Because $5x$ is positive, two positive numbers must have been multiplied. Let's try 1 and 6:

$(x + 1)(x + 6)$

Now that we've taken our first guess at the factors of $x^2 + 5x + 6$, use FOIL to test if these factors are correct:

First: $(x)(x) = x^2$
Outside: $(x)(6) = 6x$
Inside: $(1)(x) = x$
Last: $(1)(6) = 6$

$x^2 + 6x + x + 6 = x^2 + 7x + 6$, not $x^2 + 5x + 6$

The middle term came out too large. Because the factors of 6 were both multiplied by x, which has a coefficient of 1, the sum of the factors of 6 are equal to the coefficient of the middle term of $x^2 + 5x + 6$. We need two factors that multiply to 6 and add to 5. The factors 2 and 3 multiply to 6 and add to 5, so let's try those:

$(x + 2)(x + 3)$

Check with FOIL:

First: $(x)(x) = x^2$
Outside: $(x)(3) = 3x$
Inside: $(2)(x) = 2x$
Last: $(2)(3) = 6$

$x^2 + 3x + 2x + 6 = x^2 + 5x + 6$

....

TIP: Try to begin by finding the first term of each binomial. Once you have limited the possibilities of those terms, you can begin to test different factors of the constant. Those factors will be the last term of each binomial. Keep trying different combinations of factors for the last terms until you find the pair that not only multiply to the constant of the polynomial, but also yield the correct middle term of the polynomial.

....

Once we found that 2 and 3 were the factors of 6 that we needed to use as the last terms for the binomials, it did not matter if we wrote $(x + 2)(x + 3)$ or $(x + 3)(x + 2)$. 2 could have been the last term of the first binomial or the second binomial. That's because the first term in each binomial was the same. If the first term in each binomial is different, we have to be more careful about where we place the last terms.

Factoring the polynomial $2x^2 + x - 10$ is a bit tougher, but we follow the same steps. List the factors of $2x^2$ and -10:

$2x^2$: $-2, -1, 1, 2, -2x, -x, x, 2x, 2x^2$
-10: $-10, -5, -2, -1, 1, 2, 5, 10$

In order for the first term of the polynomial $2x^2 + x - 10$ to be $2x^2$, the terms $2x$ and x must be multiplied. One of the terms cannot be x^2, or else when that term is multiplied by another term that contains x, we would get an x^3 term. There is no x^3 term in $2x^2 + x - 10$. Make the first term of one binomial $2x$ and the first term of the other binomial x:

$(2x + \underline{\ \ })(x + \underline{\ \ })$

Now we need two constants that multiply to -10. The middle term of the polynomial is x, which means that the sum of the Outside and Inside products is $1x$. The Outside product is made by multiplying $2x$ by one factor of -10. The Inside product is made by multiplying x by the other factor of -10. In other words, two times one factor of -10 plus one times the other factor of -10 add to 1. Let's try a few combinations:

$$(2x + 2)(x - 5)$$

These binomials multiply to $2x^2 - 8x - 10$. The middle term is too small. Try switching the signs:

$$(2x - 2)(x + 5)$$

These binomials multiply to $2x^2 + 8x - 10$. The middle term is too large. Try switching the constants:

$$(2x - 5)(x + 2)$$

These binomials multiply to $2x^2 - x - 10$. The middle term is too small, but we are getting close. Try changing the signs:

$$(2x + 5)(x - 2)$$

These binomials multiply to $2x^2 + x - 10$. We have found the right factors.

....

TIP: It can take a little time to find the right combination of constants and signs when factoring a trinomial. Keep a list of the combinations you have tried, so that you do not try the same ones twice. Each time you try a wrong combination, try to learn from it. Ask yourself: How is the product of these factors different from the original polynomial? This can help you decide if you need to change constants or just change signs.

....

Practice 1

Factor each expression.

1. $x^2 + 12x + 35$

2. $x^2 - 6x + 8$

3. $x^2 - x - 12$

4. $2x^2 + 3x + 1$

5. $3x^2 + 8x - 11$

FACTORING THE DIFFERENCE OF PERFECT SQUARES

In Lesson 22, we saw that the product of two binomials could result in a binomial, such as when $(x + 3)$ and $(x - 3)$ are multiplied. The product of those binomials is $x^2 - 9$, which is the difference of perfect squares. Both x^2 and 9 are perfect squares, because $x^2 = (x)(x)$ and $9 = (3)(3)$. We say that $x^2 - 9$ is the difference between perfect squares because one perfect square is subtracted from the other.

To factor the difference between perfect squares, start by taking the positive square root of each term. The square root of the first term plus the square root of the second term is the first factor. The square root of the first term minus the square root of the second term is the other factor.

The binomial $4x^2 - 36$ is the difference between perfect squares, because $4x^2$ and 36 are both perfect squares. The positive square root of $4x^2$ is $2x$ and the positive square root of 36 is 6. The square root of the first term, $2x$, plus the square root of the second term, 6, is the first factor: $2x + 6$. The square root of the first term minus the square root of the second term is the second factor: $2x - 6$. Therefore, $4x^2 - 36$ factors to $(2x + 6)(2x - 6)$.

....

TIP: The first step in factoring the difference of perfect squares is to recognize that the binomial *is* the difference between perfect squares. Look for the following clues: (1) the coefficient of the variable must be a perfect square (and remember, 1 is a perfect square); (2) the exponent of the variable must be even; and (3) the constant must be a perfect square. In the last example, the coefficient of x^2, 4, is a perfect square. The exponent of x is 2, an even number, and the constant, 36, is a perfect square.

Practice 2

Factor each expression.

1. $x^2 - 16$

2. $9x^2 - 81$

3. $49x^4 - 100$

ANSWERS

Practice 1

1. To factor $x^2 + 12x + 35$, begin by listing the positive and negative factors of the first and last terms:

x^2: $-x, x$

35: $-35, -7, -5, -1, 1, 5, 7, 35$

x^2 is the square of either x or $-x$. Begin by trying x as the first term in each binomial:

$(x + \underline{})(x + \underline{})$

The coefficients of each x term are 1. The last terms of each binomial must multiply to 35 and add to 12, since $12x$ is the sum of the products of the Outside and Inside terms. $12x$ and 35 are positive, so we are looking for two positive numbers that multiply to 35 and add to 12.

$(1)(35) = 35$, but $1 + 35 = 36$

$(5)(7) = 35$, and $5 + 7 = 12$

The constant of one binomial is 5 and the constant of the other binomial is 7:

$(x + 5)(x + 7)$

Check the answer using FOIL:

First: $(x)(x) = x^2$

Outside: $(x)(7) = 7x$

Inside: $(5)(x) = 5x$

Last: $(5)(7) = 35$

$x^2 + 7x + 5x + 35 = x^2 + 12x + 35$

2. To factor $x^2 - 6x + 8$, begin by listing the positive and negative factors of the first and last terms:

$x^2: -x, x$

8: $-8, -4, -2, -1, 1, 2, 4, 8$

x^2 is the square of either x or $-x$. Begin by trying x as the first term in each binomial:

$(x + \_\_)(x + \_\_)$

The coefficients of each x term are 1. The last terms of each binomial must multiply to 8 and add to -6, since $-6x$ is the sum of the products of the Outside and Inside terms. $-6x$ is negative and 8 is positive, so we are look-ing for two negative numbers that multiply to 8 and add to -6.

$(-1)(-8) = 8$, but $-1 + (-8) = -9$

$(-2)(-4) = 8$, and $-2 + (-4) = -6$

The constant of one binomial is -2 and the constant of the other binomial is -4:

$(x - 2)(x - 4)$

Check the answer using FOIL:

First: $(x)(x) = x^2$

Outside: $(x)(-4) = -4x$

Inside: $(-2)(x) = -2x$

Last: $(-2)(-4) = 8$

$x^2 - 4x - 2x + 8 = x^2 - 6x + 8$

3. To factor $x^2 - x - 12$, begin by listing the positive and negative factors of the first and last terms:

$x^2: -x, x$

12: $-12, -6, -4, -3, -2, -1, 1, 2, 3, 4, 6, 12$

x^2 is the square of either x or $-x$. Begin by trying x as the first term in each binomial:

$(x + \_\_)(x + \_\_)$

The coefficients of each x term are 1. The last terms of each binomial must multiply to -12 and add to -1, because $-x$ is the sum of the products of the Outside and Inside terms. $-x$ is negative, as is -12, so we are looking for one positive number and one negative number that multiply to -12 and add to -1.

$(-12)(1) = -12$, but $-12 + 1 = -11$

$(12)(-1) = -12$, but $12 + (-1) = 11$

$(-6)(2) = -12$, but $-6 + 2 = -4$

$(6)(-2) = -12$, but $6 + (-2) = 4$

$(4)(-3) = -12$, but $4 + (-3) = 1$
$(-4)(3) = -12$, and $-4 + 3 = -1$

The constant of one binomial is -4 and the constant of the other binomial is 3:

$(x - 4)(x + 3)$

Check the answer using FOIL:

First: $(x)(x) = x^2$
Outside: $(x)(3) = 3x$
Inside: $(-4)(x) = -4x$
Last: $(-4)(3) = -12$
$x^2 + 3x - 4x - 12 = x^2 - x - 12$

4. To factor $2x^2 + 3x + 1$, begin by listing the positive and negative factors of the first and last terms:

$2x^2$: $-2, -1, 1, 2, -x, x$
1: $-1, 1$

$2x^2$ could be $(2x)(x)$ or $(-2x)(-x)$. Begin by trying $2x$ as the first term in one binomial and x as the first term in the other binomial:

$(2x + \_\_)(x + \_\_)$

The coefficient of one x term is 2 and the coefficient of the other x term is 1. The last terms of each binomial must multiply to 1. One of those terms will be multiplied by $2x$, and the other will be multiplied by $1x$. The sum of those products will add to $3x$. $3x$ and 1 are positive, so we are looking for two positive numbers that multiply to 1. The only possibility is 1 times 1, so the constant of each binomial is 1:

$(2x + 1)(x + 1)$

Check the answer using FOIL:

First: $(2x)(x) = 2x^2$
Outside: $(2x)(1) = 2x$
Inside: $(1)(x) = x$
Last: $(1)(1) = 1$
$2x^2 + 2x + x + 1 = 2x^2 + 3x + 1$

5. To factor $3x^2 + 8x - 11$, begin by listing the positive and negative factors of the first and last terms:

$3x^2$: $-3, -1, 1, 3, -x, x$
11: $-11, -1, 1, 11$

$3x^2$ could be $(3x)(x)$ or $(-3x)(-x)$. Begin by trying $3x$ as the first term in one binomial and x as the first term in the other binomial:

$(3x + \_\_)(x + \_\_)$

The coefficient of one x term is 3, and the coefficient of the other x term is 1. The last terms of each binomial must multiply to -11. One of those terms will be multiplied by $3x$ and the other will be multiplied by $1x$. The sum of those products will add to $8x$. $8x$ is positive, but -11 is negative, so we are looking for a positive number and a negative number that multiply to -11. The possibilities are $(-1)(11)$ and $(1)(-11)$:

$(3x - 11)(x + 1)$

Check the answer using FOIL:

First: $(3x)(x) = 3x^2$

Outside: $(3x)(1) = 3x$

Inside: $(-11)(x) = -11x$

Last: $(-11)(1) = -11$

$3x^2 + 3x - 11x - 11 = 3x^2 - 8x - 11$, not $3x^2 + 8x - 11$

Try changing the signs of the last terms of each binomial:

$(3x + 11)(x - 1)$

Check the answer using FOIL:

First: $(3x)(x) = 3x^2$

Outside: $(3x)(-1) = -3x$

Inside: $(11)(x) = 11x$

Last: $(11)(-1) = -11$

$3x^2 - 3x + 11x - 11 = 3x^2 + 8x - 11$

Therefore, $3x^2 + 8x - 11$ factors to $(3x + 11)(x - 1)$.

Practice 2

1. The binomial $x^2 - 16$ is the difference between two perfect squares. The coefficient of x, 1, is a perfect square, and the exponent of x is even. The constant, 16, is also a perfect square. The positive square root of x^2 is x and the positive square root of 16 is 4.

   The square root of the first term, x, plus the square root of the second term, 4, is the first factor: $x + 4$. The square root of the first term minus the square root of the second term is the second factor: $x - 4$. Therefore, $x^2 - 16$ factors to $(x + 4)(x - 4)$.

2. The binomial $9x^2 - 81$ is the difference between two perfect squares. The coefficient of x, 9, is a perfect square, and the exponent of x is even. The constant, 81, is also a perfect square. The positive square root of $9x^2$ is $3x$, and the positive square root of 81 is 9.

   The square root of the first term, $3x$, plus the square root of the second term, 9, is the first factor: $3x + 9$. The square root of the first term minus the square root of the second term is the second factor: $3x - 9$. Therefore, $9x^2 - 81$ factors to $(3x + 9)(3x - 9)$.

3. The binomial $49x^4 - 100$ is the difference between two perfect squares. The coefficient of x, 49, is a perfect square, and the exponent of x is even. The constant, 100, is also a perfect square. The positive square root of $49x^4$ is $7x^2$, and the positive square root of 100 is 10.

The square root of the first term, $7x^2$, plus the square root of the second term, 10, is the first factor: $7x^2 + 10$. The square root of the first term minus the square root of the second term is the second factor: $7x^2 - 10$. Therefore, $49x^4 - 100$ factors to $(7x^2 + 10)(7x^2 - 10)$.

L E S S O N 24

quadratic equation

Algebra as far as the quadratic equation and the use of logarithms are often of value in ordinary cases: but all beyond these is but a luxury; a delicious luxury indeed.

—THOMAS JEFFERSON (1743–1826)
THIRD PRESIDENT OF THE UNITED STATES

In this lesson, you'll learn how to solve a quadratic equation using factoring and the quadratic formula.

A QUADRATIC EQUATION is a quadratic expression with an equal sign. Formally, the quadratic equation is described as $ax^2 + bx + c = 0$. a is the coefficient of the x^2 term, and it cannot be 0. Otherwise there would be no x^2 term (and then the equation would not be quadratic!). b is the coefficient of the x term, and c is the constant.

Most quadratic equations have two solutions, although some have only one. To solve a quadratic equation, the first step is to get the equation in the form $ax^2 + bx + c = 0$. If the equation is equal to some value other than 0, subtract that value from both sides of the equation. That value will be combined with one of the terms on the left side of the equation, leaving the equation in the form $ax^2 + bx + c = 0$.

Once the equation is in that form, we have two ways to find the solutions. The first method is to factor the equation into two binomials. Now you know why we learned how to do that in Lesson 23. We then take the two binomials and set them each equal to 0. Why? An expression is the product of its factors.

If a value of x makes one factor equal to 0, then the product of the factors will be equal to 0, because any quantity multiplied by 0 is 0.

The equation $x^2 + 9x + 18 = 0$ is ready to be factored, because it is already in the form $ax^2 + bx + c = 0$. To factor $x^2 + 9x + 18$, we start as always by listing the factors of x^2 and the factors of 18:

x^2: $-x, x$
18: $-18, -9, -6, -3, -2, -1, 1, 2, 3, 6, 9, 18$

The middle term, $9x$ is positive, and the last term, 18, is positive, so we're looking for two positive numbers:

$(x + 1)(x + 18) = x^2 + 19x + 18$. The middle term is too large.
$(x + 2)(x + 9) = x^2 + 11x + 18$. The middle term is still too large.
$(x + 3)(x + 6) = x^2 + 9x + 18$. Got it!

The factors of $x^2 + 9x + 18$ are $(x + 3)$ and $(x + 6)$. The equation $x^2 + 9x + 18 = 0$ holds true when either $(x + 3)$ equals 0 or $(x + 6)$ equals 0. We set each factor equal to 0 and solve for x:

$x + 3 = 0$ $x + 6 = 0$

$x = -3$ $x = -6$

The roots, or the solutions of $x^2 + 9x + 18 = 0$, are $x = -3$ and $x = -6$. The **roots** of an equation are the values for which $f(x)$ is 0.

That equation had two roots, but if the trinomial in the equation comes from squaring a binomial, then the equation will have only one root.

To solve the equation $x^2 + 25 = 10x$, we must first put the equation in the form $ax^2 + bx + c = 0$. If we subtract $10x$ from both sides of the equation, we have $x^2 - 10x + 25 = 0$. Now, we are ready to factor. The factors of x^2 are $-x, x$ and the factors of 25 are $-25, -5, -1, 1, 5, 25$. Because the middle term is negative and the last term is positive, we are looking for two negative numbers that multiply to 25 and add to -10. $x^2 - 10x + 25$ factors into $(x - 5)(x - 5)$. Because $x^2 - 10x + 25 = 0$, we set each of its factors equal to 0. However, both factors are the same. They both are $(x - 5)$, so we only need to set one of them equal to 0:

$x - 5 = 0$
$x = 5$

The trinomial $x^2 - 10x + 25$ came from squaring $(x - 5)$, so the equation has only one root, 5.

Practice 1

Find the roots of each equation.

1. $x^2 + 5x - 14 = 0$

2. $x^2 - 3x - 54 = 0$

3. $5x^2 + 5x = 10$

4. $x^2 + 4x = -4$

5. $x^2 - 121 = 0$

QUADRATIC FORMULA

Not every quadratic equation can be factored easily. If the roots of a quadratic equation are not integers, it can be nearly impossible to find them by factoring. In fact, sometimes the roots are imaginary! An **imaginary number** is a number whose square is less than zero. Sounds impossible; that's why we call them imaginary numbers. There is no real number that is the square root of –4. To take the square root of a negative number, we write what the square root of the number would be if it were positive, followed by the letter i. The letter i represents the square root of –1. Because $\sqrt{-4}$ can be rewritten as $\sqrt{4}\sqrt{-1}$, we say that the square root of –4 is $2i$, which is 2 times the square root of –1.

   To find the roots of any quadratic equation, including ones with imaginary roots, we can use the quadratic formula. At first glance, the formula may look pretty complicated, but we'll break it down into pieces:

$$x = \frac{-b \pm \sqrt{b^2 - 4ac}}{2a}$$

What does it all mean? Remember, before we solve a quadratic equation, it must be in the form $ax^2 + bx + c = 0$. The coefficients a and b and the constant c each get substituted into the quadratic formula. Then, evaluate the expression on the right side of the equation to find the values of x.

   First, let's look at a quadratic equation that we've already solved. We know that the solutions of $x^2 + 9x + 18 = 0$ are –3 and –6, but we can find those same roots using this formula. In the equation $x^2 + 9x + 18 = 0$, a is the coefficient of x^2, which is 1. b is the coefficient of x, which is 9, and c is the constant, which is 18. Now, substitute these values into the formula:

$$x = \frac{-9 \pm \sqrt{(9)^2 - 4(1)(18)}}{2(1)}$$

Start with the values under the square root sign. Square 9 and multiply 4 by 1 and 18:

$$\frac{-9 \pm \sqrt{81 - 72}}{2(1)}$$

Next, subtract 72 from 81. In the denominator, multiply 2 by 1:

$$\frac{-9 \pm \sqrt{9}}{2}$$

Take the square root of 9:

$$\frac{-9 \pm 3}{2}$$

The symbol between –9 and 3 is the plus-minus symbol. Remember: quadratic equations usually have two roots. One root is equal to –9 plus 3, divided by 2, and the other root is equal to –9 minus 3, divided by 2:

$$\frac{-9 + 3}{2} , \frac{-9 - 3}{2}$$

The first fraction is equal to $-6 \div 2 = -3$, and the second fraction is equal to $-12 \div 2 = -6$. The solutions of the equation are –3 and –6, the same roots we found by factoring.

Now that we know how to use the formula, let's try it with a tougher example: $2x^2 - 6x + 7 = 0$. The coefficient of x^2 is 2, so $a = 2$. The coefficient of x is –6, so $b = -6$, and the constant is 7, so $c = 7$. Substitute these values into the quadratic formula:

$$x = \frac{6 \pm \sqrt{(-6)^2 - 4(2)(7)}}{2(2)}$$

This simplifies to:

$$x = \frac{6 \pm \sqrt{-20}}{4}$$

The square root of -20 is imaginary. It can be rewritten as $\sqrt{4}\sqrt{5}\sqrt{-1}$. The square root of 4 is 2 and the square root of -1 is i. Therefore, write the square root of -20 as $2\sqrt{5}\,i$.

We now have:

$$x = \frac{6 \pm 2\sqrt{5}i}{4}$$

One root is equal to $\frac{6 + 2\sqrt{5}i}{4}$ and simplifies to $\frac{3 + \sqrt{5}i}{2}$, and the other is equal to $\frac{6 - 2\sqrt{5}i}{4}$ and simplifies to $\frac{3 - \sqrt{5}i}{2}$.

....

TIP: Every quadratic equation can be solved using the quadratic formula, but it is often easier to use factoring to the find the roots of the equation if the roots are integers. Before using the quadratic formula, try to factor the trinomial. If factoring looks too difficult, then use the quadratic formula.

....

Practice 2

Find the solutions to each equation.

1. $x^2 - 5x + 3 = 0$

2. $x^2 + 2x + 2 = 0$

ANSWERS

Practice 1

1. The equation $x^2 + 5x - 14 = 0$ is in the form $ax^2 + bx + c = 0$. Factor $x^2 + 5x - 14$ and set each factor equal to 0.

Factors of x^2: $-x, x$

Factors of 14: $-14, -7, -2, -1, 1, 2, 7, 14$

Because the last term in the equation is –14, look for a negative number and a positive number that multiply to –14:

$(x – 1)(x + 14) = x^2 + 13x – 14$. The middle term is too large.

$(x – 7)(x + 2) = x^2 – 5x – 14$. The middle term is the wrong sign.

$(x – 2)(x + 7) = x^2 + 5x – 14$. These are the correct factors.

Set $x – 2$ and $x + 7$ equal to 0 and solve for x:

$x – 2 = 0$ $\qquad\qquad\qquad$ $x + 7 = 0$

$x = 2$ $\qquad\qquad\qquad\quad$ $x = –7$

The solutions of $x^2 + 5x – 14 = 0$ are 2 and –7.

2. The equation $x^2 – 3x – 54 = 0$ is in the form $ax^2 + bx + c = 0$. Factor $x^2 – 3x – 54$ and set each factor equal to 0.

Factors of x^2: $–x, x$

Factors of 54: –54, –27, –18, –9, –6, –3, –2, –1, 1, 2, 3, 6, 9, 18, 27, 54

Because the last term in the equation is –54, look for a negative number and a positive number that multiply to –54:

$(x – 54)(x + 1) = x^2 – 53x – 54$. The middle term is too small.

$(x – 27)(x + 2) = x^2 – 25x – 54$. The middle term is still too small.

$(x – 18)(x + 3) = x^2 – 15x – 54$. The middle term is still too small.

$(x – 9)(x + 6) = x^2 – 3x – 54$. These are the correct factors.

Set $x – 9$ and $x + 6$ equal to 0 and solve for x:

$x – 9 = 0$ $\qquad\qquad\qquad$ $x + 6 = 0$

$x = 9$ $\qquad\qquad\qquad\quad$ $x = –6$

The solutions of $x^2 – 3x – 54 = 0$ are 9 and –6.

3. The equation $5x^2 + 5x = 10$ is not in the form $ax^2 + bx + c = 0$. Subtract 10 from both sides of the equation. The equation is now $5x^2 + 5x – 10 = 0$. Factor $5x^2 + 5x – 10$ and set each factor equal to 0.

Factors of $5x^2$: $–5, –1, 1, 5, –x, x$

Factors of –10: –10, –5, –2, –1, 1, 2, 5, 10

Because the last term is –10, look for a negative number and a positive number that multiply to –10:

$(5x – 10)(x + 1) = 5x^2 – 5x – 10$. The middle term is too small.

$(5x – 5)(x + 2) = 5x^2 + 5x – 10$. These are the correct factors.

Set $5x – 5$ and $x + 2$ equal to 0 and solve for x:

$5x – 5 = 0$ $\qquad\qquad\qquad$ $x + 2 = 0$

$5x = 5$ $\qquad\qquad\qquad\quad$ $x = –2$

$x = 1$

The solutions of $5x^2 + 5x – 10 = 0$ are 1 and –2.

4. The equation $x^2 + 4x = -4$ is not in the form $ax^2 + bx + c = 0$. Add 4 to both sides of the equation. The equation is now $x^2 + 4x + 4 = 0$. Factor $x^2 + 4x + 4$ and set each factor equal to 0.

Factors of x^2: $-x, x$

Factors of 4: $-4, -2, -1, 1, 2, 4$

Because the last term of $x^2 + 4x + 4$ is 4 and the middle term is positive, look for two positive numbers that multiply to 4:

$(x + 1)(x + 4) = x^2 + 5x + 4$. The middle term is too large.

$(x + 2)(x + 2) = x^2 + 4x + 4$. These are the correct factors.

Both factors are the same, so set just $x + 2$ equal to 0 and solve for x:

$x + 2 = 0$

$x = -2$

The only solution of $x^2 + 4x + 4 = 0$ is -2.

5. The equation $x^2 - 121 = 0$ is in the form $ax^2 + bx + c = 0$. x^2 and 121 are both perfect squares, and $x^2 - 121$ is the difference of two squares. The positive square root of x^2 is x and the positive square root of 121 is 11.

The square root of the first term, x, plus the square root of the second term, 11, is the first factor: $x + 11$. The square root of the first term minus the square root of the second term is the second factor: $x - 11$. $x^2 - 121$ factors to $(x + 11)(x - 11)$.

Set $x + 11$ and $x - 11$ equal to 0 and solve for x:

$x + 11 = 0$ $\qquad\qquad$ $x - 11 = 0$

$x = -11$ $\qquad\qquad$ $x = 11$

The solutions of $x^2 - 121 = 0$ are -11 and 11.

Practice 2

1. In the equation $x^2 - 5x + 3 = 0$, the coefficient of x^2 is 1, so $a = 1$. The coefficient of x is -5, so $b = -5$, and the constant is 3, so $c = 3$. Substitute these values into the quadratic formula:

$$x = \frac{5 \pm \sqrt{(-5)^2 - 4(1)(3)}}{2(1)}$$

$(-5)^2 = 25$ and $(4)(1)(3) = 12$:

$$x = \frac{5 \pm \sqrt{25 - 12}}{2(1)}$$

Under the square root symbol, $25 - 12 = 13$, and in the denominator, $2(1) = 2$:

$$x = \frac{5 \pm \sqrt{13}}{2}$$

The square root of 13 cannot be simplified. One solution is equal to $\frac{5+\sqrt{13}}{2}$ and the other is equal to $\frac{5-\sqrt{13}}{2}$.

2. In the equation $x^2 + 2x + 2 = 0$, the coefficient of x^2 is 1, so $a = 1$. The coefficient of x is 2, so $b = 2$, and the constant is 2, so $c = 2$. Substitute these values into the quadratic formula:

$$x = \frac{-2 \pm \sqrt{(2)^2 - 4(1)(2)}}{2(1)}$$

$2^2 = 4$ and $(4)(1)(2) = 8$

$$x = \frac{-2 \pm \sqrt{4 - 8}}{2(1)}$$

Under the square root symbol, $4 - 8 = -4$, and in the denominator, $2(1) = 2$:

$$x = \frac{-2 \pm \sqrt{-4}}{2}$$

The square root of -4 is imaginary. It can be rewritten as $\sqrt{4}\sqrt{-1}$. The square root of 4 is 2 and the square root of -1 is i. The square root of -4 is $2i$.

$$x = \frac{-2 \pm 2i}{2}$$

One solution is equal to $\frac{-2+2i}{2}$, which simplifies to $-1 + i$, and the other is equal to $\frac{-2-2i}{2}$, which simplifies to $-1 - i$.

S E C T I O N 3

using algebra

MANY OF THE problems we have seen through the first two sections of this book have been pure algebra. We have solved an equation for x or simplified an inequality that contains x and y. But x and y have just been variables to us. Now, we will encounter some real-world situations that require us to use algebra. What is the probability of drawing two aces from a deck of cards? How much interest will I earn if I keep $200 in a bank for one year? If my boss gives me a 10% raise, how much money will I be making? In the next few lessons, you will see how the value of x is not always just some number. It could be the answer to a real question of your own.

This section will show you how to apply your new algebra skills to real-life problems, including:

- word problems
- ratios and proportions
- statistics problems
- probability problems
- percent increase and percent decrease
- simple interest
- arithmetic and geometric sequences
- perimeter, area, and volume

algebra word problems

There is no branch of mathematics, however abstract, which may not some day be applied to phenomena of the real world.
—NIKOLAI LOBATCHEVSKY (1792–1856)
RUSSIAN MATHEMATICIAN

In this lesson, you'll learn how to use algebra to solve real-life problems.

YOU MIGHT NOT REALIZE how often algebra can be used to answer everyday questions. Anytime a situation has an unknown value, that unknown can be found using algebra. If a gas tank holds 20 gallons of gas, and a car goes 16 miles for every gallon, how much gas will be left in the tank after a trip of 300 miles? That's an important question to answer if you do not want to run out of gas!

To solve a word problem, we follow these six steps:

1. Read the entire word problem.
2. Underline the keywords.
3. List the possible operations.
4. Represent the unknown with a variable.
5. Write an equation or inequality to solve the problem.
6. Solve the number sentence.

The first step might seem simple, but it is actually the most important. If you do not read a problem carefully, you might not answer the question that is being asked. Next, underline the keywords that might signal which operations must be performed. Remember the chart from Lesson 6?

Addition	Subtraction	Multiplication	Division
combine	take away	times	share
together	difference	product	quotient
total	minus	factor	percent
plus	decrease	each	out of
sum	left	every	average
altogether	more than	increase	each
and	fewer	per	per
increase	less than		

We used this chart to help us turn algebraic phrases into algebraic expressions. We can also use it to help us turn algebraic word problems into algebraic equations or inequalities. Once we know what operation or operations are needed, we use a variable to hold the place of the unknown and write an equation or inequality to solve the problem. Finally, we solve that for the variable and check our work.

Example
Bethel has five less than four times the number of postcards that Loring has. If Bethel has 11 postcards, how many postcards does Loring have?

First, read the entire problem. We are told how many postcards Loring has and we are looking for how many postcards Bethel has. Next, underline the keywords. The words *less than* can signal subtraction and the word *times* can signal multiplication. We don't know how many postcards Loring has, so we can use x to represent that value.

Now, we must write an equation that we can use to solve this problem. Bethel has five less than four times the number of postcards that Loring has. The number of postcards Loring has is x, so Bethel has five less than four times x. *Four times x* means "four multiplied by x," which is $4x$. Five less than $4x$ is $4x - 5$. The expression $4x - 5$ represents how many postcards Bethel has. We are told that Bethel has 11 postcards, so we can set these two values equal to each other: $4x - 5 = 11$. Finally, solve the equation for x to find how many postcards Loring has:

$4x - 5 = 11$
$4x = 16$
$x = 4$

Loring has 4 postcards. We can check our work by rereading the word problem. Bethel has five less than four times the number of postcards that Loring has. Four times four is 16, and five less than that is 11, which is the number of postcards Bethel has. We have found the correct answer.

....

TIP: If you are not sure which operation should be used to solve a problem, write more than one equation, using different operations in each equation. Solve each equation and decide which answers seem reasonable. Then, check each answer against the information given in the question and choose the one that correctly solves the problem.

....

Example
Kerrie places a hose in a wading pool that holds 70 gallons of water. If the pool contains 15 gallons before the hose is turned on, and the hose pours water at a rate of 5 gallons per minute, how long will it take to fill the pool?

We are given the volume of the pool, 70 gallons, its starting volume, 15 gallons, and the rate at which the pool fills, 5 gallons per minute. The keyword *per* can signal multiplication or division. Let's write an equation using multiplication and another using division and decide which equation makes more sense. We can use x to represent the number of minutes it will take to fill the pool.

The total volume, 70, is equal to 15 gallons plus the amount of water added by the hose, which is either five times x or five divided by x:

$5x + 15 = 70$
$\frac{x}{5} + 15 = 70$

Solve each equation for x:

$5x + 15 = 70$

$5x = 55$

$x = 11$

$\frac{x}{5} + 15 = 70$

$\frac{x}{5} = 55$

$x = 275$

The answer to the first equation is 11 minutes and the answer to the second equation is 275 minutes, which is more than 4 hours. An answer of 11 minutes is more reasonable. If 5 gallons are added to a pool every minute, then the number of gallons added in one minute would be equal to 1 times 5, not 1 divided by 5. The number of gallons added in 11 minutes would be equal to 11 times 5, which is 55. Since the pool already contained 15 gallons, after 11 minutes, the pool would be filled with 70 gallons of water.

INEQUALITY WORD PROBLEMS

You can find algebraic inequalities in everyday situations, too. The same set of steps can be used to find the answers to these problems.

Example
Jerome has $25. He buys a hot dog for $1 and then rides the Ferris wheel. If each ride costs $4, how many rides can Jerome take?

Read the entire problem. We are told how much money Jerome has, how much a hot dog costs, and how much a ride on the Ferris wheel costs. Jerome eats one hot dog, and we are looking for the number of rides that Jerome can take.

Next, underline the keywords. The word *each* can signal multiplication or division. Since we are given the cost of one ride, that number can be multiplied by the number of rides to find the total cost of the rides. We are also told that Jerome buys a hot dog, which means that the amount of money he has decreases. We do not know how many rides Jerome can take, so we can use x to represent that value.

We are looking for how many rides Jerome can take, which is a range of values, not a single value. There is a maximum number of rides that Jerome can take, but he could also take fewer rides than that number. We will write an inequality to solve this problem.

Jerome has $25 and spends $1 on a hot dog. We must subtract 1 from 25 to find how much money Jerome has left. Because each ride costs $4, the total cost of the rides is equal to four times x, which must be less than or equal to the amount of money Jerome has left. The inequality $4x \leq 25 - 1$ can be solved for x, the number of rides Jerome can take:

$$4x \leq 25 - 1$$
$$4x \leq 24$$
$$x \leq 6$$

Jerome can take as many as 6 rides on the Ferris wheel, but no more than that. He could also take fewer than 6 rides.

Practice

Find the answer to each question.

1. Every week, Natalie swims three more than six times the number of laps Michael swims. If Natalie swims 45 laps, how many laps does Michael swim?

2. Josh fills 60 cups of water from a 500-ounce water cooler. If there are 20 ounces left in the cooler when Josh has finished, and each cup contains the same amount of water, how much water is in each cup?

3. Tanya has $25 in her bank. She adds to the bank one-fifth of the money she receives for her birthday, bringing the total to $41. How much money did Tanya receive for her birthday?

4. A gym has a membership fee of $50 and a monthly fee of $30. If Angela has $200, for how many months could she belong to the gym?

5. Tatiana needs $150 to go to camp. She has $38 saved, and she sells raffle tickets for $4 each. How many tickets must she sell in order to go to camp?

ANSWERS

Practice

1. We are looking for the number of laps Michael swims, so we can use x to represent that number. Natalie swims three more than six times that number, which means that she swims $6x + 3$ laps. We know that she swims 45 laps, so we can set $6x + 3$ equal to 45 and solve for x:
    $$6x + 3 = 45$$
    $$6x = 42$$
    $$x = 7$$
    Michael swims 7 laps.

2. We're looking for the volume of water in each cup, so we can use x to represent that number. Josh fills 60 cups with water, so 60 times x, or $60x$, is equal to the amount of water used to fill the cups. The cooler has 20 ounces left after the cups are filled, which means that the amount of water in the cups, $60x$, plus 20 is equal to 500:
    $$60x + 20 = 500$$
    $$60x = 480$$
    $$x = 8$$
    Each cup of water contains 8 ounces.

3. We're looking for the amount of money Tanya received for her birthday, so we can use x to represent that number. She adds to the bank one-fifth of what she received. Divide what she received, x, by 5: $\frac{x}{5}$. She had $25 in her bank already, so add 25 to $\frac{x}{5}$. The total is equal to the total in her bank now, $41:
    $$\frac{x}{5} + 25 = 41$$
    $$\frac{x}{5} = 16$$
    $$x = 80$$
    Tanya received $80 for her birthday.

4. We're looking for the number of months Angela could belong to the gym, so use x to represent that number. There is a maximum number of months Angela could belong to the gym, but she could also belong to the gym for a shorter period of time, so we need an inequality. The monthly fee is $30, so the amount of money spent in monthly fees is 30 times the number of months, or $30x$. Angela must also pay the membership fee, $50, and the

total of the monthly fee and the membership fee must be less than or equal to the amount of money Angela has, $200:

$$30x + 50 \leq 200$$
$$30x \leq 150$$
$$x \leq 5$$

Angela can belong to the gym for 5 or fewer months.

5. We're looking for the number of tickets Tatiana must sell, so we can use x to represent that number. There is a minimum number of tickets Tatiana must sell, but she could also sell more than that number, so we need an inequality. Each raffle ticket costs $4, so the amount of money Tatiana collects is equal to 4 times the number of tickets she sells, or $4x$. She already has $38, so the total money collected from selling tickets plus that $38 must be greater than or equal to the cost of going to camp, $150:

$$4x + 38 \geq 150$$
$$4x \geq 112$$
$$x \geq 28$$

Tatiana must sell 28 or more tickets to go to camp.

using algebra: ratios and proportions

A man is like a fraction whose numerator is what he is and whose denominator is what he thinks of himself. The larger the denominator the smaller the fraction.
—LEV NIKOLAYEVICH TOLSTOY (1828–1920)
RUSSIAN WRITER

In this lesson, you'll learn how to use algebra to define ratios and proportions and find exact values using these ratios and proportions.

A RATIO IS A RELATIONSHIP between two or more quantities. If there are 15 girls and 10 boys in your class, we can say that the ratio of girls to boys is "15 to 10." That ratio can be written with a colon, as in 15:10, or as a fraction, as in $\frac{15}{10}$. Fractions represent division, and we can simplify ratios just as we simplify fractions. Because the greatest common factor of 15 and 10 is 5, the ratio $\frac{15}{10}$ reduces to $\frac{3}{2}$. For every 3 girls in your class, there are 2 boys.

A **proportion** is an equation that shows two equal ratios. Because the ratio $\frac{15}{10}$ reduces to $\frac{3}{2}$, we can write the proportion $\frac{15}{10} = \frac{3}{2}$. We say that these two ratios are *in proportion* to each other.

Example
The ratio of apples to oranges in a basket is 5:2. If there are 25 apples in the basket, how many oranges are in the basket?

We are given the ratio of apples to oranges, which we can write as the fraction $\frac{5}{2}$. Because we know that there are 25 apples, the ratio $\frac{5}{2}$ is a reduced form of the ratio of the number of actual apples to the number of actual oranges. We do not know the number of actual oranges, so we will represent that with x. The ratio of the number of actual apples to the number of actual oranges is $\frac{25}{x}$. Now that we have two equal ratios, we can write a proportion that sets them equal to each other: $\frac{5}{2} = \frac{25}{x}$. To solve a proportion, we cross multiply: Multiply the numerator of the first fraction by the denominator of the second fraction, and multiply the numerator of the second fraction by the denominator of the first fraction. Then, set those two products equal to each other:

$\frac{5}{2} = \frac{25}{x}$

$(5)(x) = (2)(25)$

$5x = 50$

$x = 10$

The number of actual oranges is 10. We can check our answer by putting 10 into the ratio of actual apples to actual oranges. Because $\frac{25}{10}$ reduces to $\frac{5}{2}$, 10 must be the correct answer.

....

TIP: The order of a ratio is very important. The ratio 5:2 is not the same as the ratio 2:5. If we say that there are five apples for every two oranges, we must write 5:2. The ratio 2:5 would mean that there are two apples for every five oranges, which is a different relationship.

....

We can also use a proportion to find an exact number, given a ratio and a total. If the ratio of apples to oranges is 2:3, and the total number of pieces of fruit is 40, we can write a part-to-total ratio. There are 2 apples for every 3 oranges, which means that there are 2 apples for every $2 + 3 = 5$ pieces of fruit. We can find the actual number of apples by comparing the ratio of apples to total pieces of fruit, 2:5, to the ratio of actual apples to actual total number of pieces of fruit, x:40. Now, we can write a proportion and solve for x:

$\frac{2}{5} = \frac{x}{40}$

$(5)(x) = (2)(40)$

$5x = 80$

$x = 16$

If there are 40 pieces of fruit and the ratio of apples to oranges is 2:3, then there are 16 apples in the basket.

Practice 1

Answer the following questions.

1. The ratio of red cars to blue cars in a parking lot is 1:6. If there are 36 blue cars, how many red cars are in the lot?

2. The ratio of lemon candies to strawberry candies in a dish is 3:8. If there are 9 lemon candies, how many strawberry candies are in the dish?

3. If the ratio of bass to trout in a pond is 4:5, and there are 135 trout, how many bass are in the pond?

4. If the ratio of basketball players to football players at Roward Middle School is 2:7, and there are 54 total basketball and football players, how many basketball players are there?

5. The ratio of adults to children at a movie is 3:4. If there are 196 people at the movie, how many of them are children?

ALGEBRAIC EXPRESSIONS IN RATIOS

The values in a ratio can also be algebraic expressions. We can use ratios and proportions to solve for the value of the variable in these expressions.

Example
The ratio of sopranos to altos in a choir is 5:7. If there are 6 more altos than sopranos in the choir, how many sopranos are in the choir?

The number of sopranos is unknown, so we will let x represent that number. There are 6 more altos than sopranos, which means that the number of altos is equal to $x + 6$. The ratio of sopranos to altos is 5:7, or $\frac{5}{7}$. We can write the ratio of actual sopranos, x, to actual altos, $x + 6$, as a ratio, too: $\frac{x}{x+6}$. Since these ratios are equal, we can write a proportion and solve for x:

$\frac{5}{7} = \frac{x}{x+6}$
$7x = 5x + 30$
$2x = 30$
$x = 15$

There are 15 sopranos in the choir.

TIP: There is more than one way to write a ratio that contains algebraic expressions. In the previous example, if we had wanted to find the number of altos in the choir, we could have let x equal the number of altos, and the number of sopranos would have been $x - 6$, since there are 6 more altos than sopranos. We would have set $\frac{5}{7}$ equal to $\frac{x-6}{x}$, and found that x, the number of altos, is equal to 21. As always, let x represent the value that you are looking to find.

Practice 2

Answer the following questions.

1. The ratio of black pens to blue pens in a box is 7:4. If there are 9 more black pens than blue pens, how many blue pens are there?

2. The ratio of fiction books to nonfiction books on a shelf is 6:5. If there are 4 more fiction books than nonfiction books, how many fiction books are there?

3. The ratio of parents to students on a school trip is 2:9. If the number of students was 6 more than 4 times the number of parents, how many parents went on the trip?

ANSWERS

Practice 1

1. Write the ratio of red cars to blue cars as a fraction. $1:6 = \frac{1}{6}$.
   The number of red cars is unknown, so represent that number with x.
   The ratio of actual red cars to actual blue cars is $x:36$, or $\frac{x}{36}$.
   Set these ratios equal to each other, cross multiply, and solve for x:
   $$\frac{1}{6} = \frac{x}{36}$$
   $$6x = 36$$
   $$x = 6$$
   There are 6 red cars in the parking lot.

2. Write the ratio of lemon candies to strawberry candies as a fraction. $3:8 = \frac{3}{8}$. The number of strawberry candies is unknown, so represent that number with x. The ratio of actual lemon candies to actual strawberry candies is $9:x$, or $\frac{9}{x}$.

Set these ratios equal to each other, cross multiply, and solve for x:
$$\frac{3}{8} = \frac{9}{x}$$
$$3x = 72$$
$$x = 24$$

There are 24 strawberry candies in the dish.

3. Write the ratio of bass to trout as a fraction. $4:5 = \frac{4}{5}$. The number of bass is unknown, so represent that number with x. The ratio of actual bass to actual trout is $x:135$, or $\frac{x}{135}$.

Set these ratios equal to each other, cross multiply, and solve for x:
$$\frac{4}{5} = \frac{x}{135}$$
$$5x = 540$$
$$x = 108$$

There are 108 bass in the pond.

4. The ratio of basketball players to football players is $2:7$, which means that the ratio of basketball players to total players is $2:(2 + 7) = 2:9$. Write this ratio as a fraction: $\frac{2}{9}$.

The number of basketball players is unknown, so represent that number with x. The ratio of actual basketball players to actual total players is $x:54$, or $\frac{x}{54}$.

Set these ratios equal to each other, cross multiply, and solve for x:
$$\frac{2}{9} = \frac{x}{54}$$
$$9x = 108$$
$$x = 12$$

There are 12 basketball players at Roward Middle School.

5. The ratio of adults to children is $3:4$, which means that the ratio of children to total people is $4:(3 + 4) = 4:7$. Write this ratio as a fraction. $4:7 = \frac{4}{7}$. The number of children is unknown, so represent that number with x. The ratio of actual children to actual people is $x:196$, or $\frac{x}{196}$.

Set these ratios equal to each other, cross multiply, and solve for x:
$$\frac{4}{7} = \frac{x}{196}$$
$$7x = 784$$
$$x = 112$$

There are 112 children at the movie.

Practice 2

1. The number of blue pens is unknown, so represent that number with x. There are 9 more black pens than blue pens, which means that the number of black pens is equal to $x + 9$. The ratio of black pens to blue pens is 7:4, or $\frac{7}{4}$. The ratio of actual black pens to actual blue pens is $(x + 9):x$, or $\frac{x+9}{x}$. These ratios are equal. Write a proportion and solve for x:

$$\frac{7}{4} = \frac{x+9}{x}$$
$$7x = 4x + 36$$
$$3x = 36$$
$$x = 12$$

There are 12 blue pens in the box.

2. The number of fiction books is unknown, so represent that number with x. There are 4 more fiction books than nonfiction books, which means that the number of nonfiction books is equal to $x - 4$. The ratio of fiction books to nonfiction books is 6:5, or $\frac{6}{5}$. The ratio of actual fiction books to actual nonfiction books is $x:(x - 4)$, or $\frac{x}{x-4}$. These ratios are equal. Write a proportion and solve for x:

$$\frac{6}{5} = \frac{x}{x-4}$$
$$5x = 6x - 24$$
$$-x = -24$$
$$x = 24$$

There are 24 fiction books on the shelf.

3. The number of parents is unknown, so represent that number with x. The number of students is 6 more than 4 times the number of parents, which means that the number of students is equal to $4x + 6$. The ratio of parents to students is 2:9, or $\frac{2}{9}$. The ratio of actual parents to actual students is $x:(4x + 6)$, or $\frac{x}{4x+6}$. These ratios are equal. Write a proportion and solve for x:

$$\frac{2}{9} = \frac{x}{4x+6}$$
$$9x = 8x + 12$$
$$x = 12$$

Therefore, 12 parents went on the trip.

using algebra:
statistics and probability

42.7% of all statistics are made up on the spot.
—STEVEN WRIGHT (1955–   )
AMERICAN COMEDIAN

In this lesson, you'll learn how to use algebra to solve statistics and probability problems, such as finding the missing value in a data set or how a probability changes with the addition of more data.

ALGEBRA CAN BE USED in various fields of mathematics. Anytime you have an unknown value, you can represent it with x (or any other letter).

In statistics, data can be analyzed by finding the mean, median, mode, and range of a set. Algebra can help us find the mean and the range of a set as new values are added to the set, and it can help us find a missing value in a set.

The **mean** of a set can be found by dividing the sum of all the values in the set by the number of values in the set. For example, the mean of the set {1, 1, 2, 5, 7, 8} is 4, because the sum of the values in the set is 24, the number of values in the set is 6, and $24 \div 6 = 4$.

We did not need algebra to tell us that, but what if we were told that a seventh number was added to the set that changed the mean to 5? How could we find that added number? Because we don't know what the number is, we represent it with x, and add x to the set: {1, 1, 2, 5, 7, 8, x}. We find the mean by adding the values in the set and dividing by the number of values, and even though our set now contains a variable, we can still do that: $1 + 1 + 2 + 5 + 7 + 8 + x = 24 + x$. We are told that the new mean is 5. We now have seven values, so we know that the sum, $24 + x$, divided by 7 is 5. We can write an equation and solve for x:

$$\frac{24 + x}{7} = 5$$
$$24 + x = 35$$
$$x = 11$$

The number that was added to the set was 11. The set is now {1, 1, 2, 5, 7, 8, 11}.

FINDING A MEDIAN

The **median** of a data set is the value that is in the middle of the set after the set is put in order from least to greatest. If the set has an even number of values, then the median is the average of the two middle values. In the set {1, 3, 5, 5, 7}, 5 is the median, because it is the middle value.

We are told that a new value is added to the set, and the median is now 4. What could have been that new value? How can we represent the set of numbers now? If 4 had been added to the set, we would have {1, 3, 4, 5, 5, 7}, and the median would have been the average of 4 and 5, which is 4.5. Because the new median is 4, not 4.5, the new value could not have been 4. Therefore, the median must be the average of two numbers, neither of which is 4. Look at the set again: {1, 3, 5, 5, 7}. If the new value is 5 or greater, the median will still be 5. However, if the new value is 3, the set becomes {1, 3, 3, 5, 5, 7}, and the median is the average of 3 and 5, which is 4. The new value could be 3, but hold on. If the new value is less than 3, the median will still be 4, because the two middle values in the set will still be 3 and 5. Let's represent the new value with x. The new value can be any number less than or equal to 3: $x \leq 3$.

FINDING A RANGE

The **range** of a data set is the difference between its greatest value and its least value. In the set {1, 3, 5, 5, 7}, the range is 6, because the greatest value is 7, the least value is 1, and $7 - 1 = 6$.

Now we are told that a new value has been added to the set, and the range is now 10. Use x to represent the new value. What could that value be? Because the range has changed, the new value is now either the smallest value in the set or the largest value in the set. If the new value is the smallest value, then the range, 10, is equal to $7 - x$. If the new value is the largest value, then the range is equal to $x - 1$:

$7 - x = 10$

$-x = 3$

$x = -3$

$x - 1 = 10$

$x = 11$

The new value is either -3 or 11.

Practice 1

Find the answer to each question.

1. The mean of a set of six values is 8. If five of the values are 4, 6, 7, 10, and 12, what is the sixth value?

2. What value must be added to the set $\{-3, -2, 5, 7, 8, 8, 12\}$ to make the mean 4?

3. A set contains the values $\{10, 11, 14, 18, 20, 21\}$. A seventh value is added. What values could be added to the set without changing the median of the set to either 14 or 18?

4. What values could be added to the set $\{2, 6, 6, 6, 12, 15\}$ to make the range 15?

PROBABILITY

Probability is the likelihood that an event or events will occur. We usually write probability as a fraction. The denominator is the total number of possibilities, and the numerator is the number of possibilities that make an event true. For example, a coin has two sides, heads and tails, so the probability of a coin landing on heads is $\frac{1}{2}$, because there are two possibilities (heads or tails) and only one possibility that makes the event "landing on heads" true.

Example
A gumball machine contains 12 red gumballs, 8 green gumballs, 5 yellow gumballs, and an unknown number of orange gumballs. If the probability of selecting a green gumball is $\frac{1}{4}$, how many orange gumballs are in the machine?

The probability of selecting a green gumball is $\frac{1}{4}$. We know that this probability is equal to the number of green gumballs, 8, divided by the total number of gumballs. We can let x represent the number of orange gumballs. The total number of gumballs is equal to $12 + 8 + 5 + x = 25 + x$. The fraction $\frac{1}{4}$ is equal to $\frac{8}{25+x}$, because both fractions represent the probability of selecting a green gumball. When two fractions equal each other, we can cross multiply: Multiply the numerator of the first fraction by the denominator of the second fraction, and multiply the numerator of the second fraction by the denominator of the first fraction. Then, set those products equal to each other:

$$\frac{1}{4} = \frac{8}{25+x}$$
$$25 + x = 32$$

Now we have an equation we can solve. By subtracting 25 from both sides, we find that $x = 7$. There are 7 orange gumballs in the machine.

....

TIP: When a new value is added to a data set, that new value will likely change the denominator of probabilities for the data set, and you will have to add x to the denominator of these probabilities. If you are looking for the probability of selecting that new value, you may have to add x to the numerator of these probabilities, too.

....

Example
The gumball machine contains 12 red gumballs, 8 green gumballs, 5 yellow gumballs, and 7 orange gumballs. How many green gumballs must be added to make the probability of selecting a green gumball $\frac{1}{2}$?

The probability of selecting a green gumball is equal to the number of green gumballs divided by the total number of gumballs. Let x represent the number of green gumballs added to the machine. That means there are now $8 + x$ green gumballs, and $12 + 8 + 5 + 7 + x = 32 + x$ total gumballs. The probability of selecting a green gumball is equal to $\frac{8+x}{32+x}$. The value of x will make this probability equal to $\frac{1}{2}$, so set the two fractions equal to each other and solve for x:

$$\frac{1}{2} = \frac{8+x}{32+x}$$

Cross multiply:

$$2(8 + x) = 32 + x$$
$$16 + 2x = 32 + x$$
$$16 + x = 32$$
$$x = 16$$

For the probability of selecting a green gumball to become $\frac{1}{2}$, 16 green gumballs must be added to the gumball machine.

Practice 2

1. A deck of cards contains 7 hearts, 9 spades, 11 clubs, and an unknown number of diamonds. If the probability of selecting a spade is $\frac{3}{11}$, how many diamonds are in the deck?

2. A sack contains 4 baseballs, 5 golf balls, and 6 tennis balls. How many baseballs must be added to the sack to make the probability of selecting a baseball $\frac{3}{4}$?

3. A jar holds 14 pennies, 16 nickels, 6 dimes, and 12 quarters. How many nickels must be removed from the jar to make the probability of selecting a nickel $\frac{1}{5}$?

ANSWERS

Practice 1

1. The mean of a set is equal to the sum of the values divided by the number of values. Let x represent the sixth value in the set:

$$4 + 6 + 7 + 10 + 12 + x = 39 + x$$
$$\frac{39 + x}{6} = 8$$
$$39 + x = 48$$
$$x = 9$$

The sixth value of the set is 9.

2. The mean of a set is equal to the sum of the values divided by the number of values. The set contains seven values, so let x represent the eighth value:

$$-3 + -2 + 5 + 7 + 8 + 8 + 12 + x = 35 + x$$
$$\frac{35 + x}{8} = 4$$
$$35 + x = 32$$
$$x = -3$$

The eighth value of the set is –3.

3. The median value of a set is the middle value of the set after the values have been ordered from least to greatest. The set {10, 11, 14, 18, 20, 21} has an even number of values, so the median is the average of the two middle values, 14 and 18. $14 + 18 = 32$, and $32 \div 2 = 16$. The median of the set is 16. When a new value is added to the set, there will be an odd number of values in the set, and after the values are ordered, the median will be the fourth value. If the number is 14 or less, 14 will become the median. If the median is 18 or greater, 18 will become the median. The only values that could be added without the median becoming 14 or 18 are values that are between 14 and 18. If x is the new value, then $14 < x < 18$.

4. The range of a data set is the difference between its greatest value and its least value. In the set {2, 6, 6, 6, 12, 15}, the greatest value is 15 and the least value is 2. For the range to change, the new value must be either less than 2 or greater than 15. Let x represent the new value in the set. Because the new range is 15, either $15 - x = 15$ or $x - 2 = 15$:

$$15 - x = 15 \qquad\qquad x - 2 = 15$$
$$-x = 0 \qquad\qquad\qquad x = 17$$
$$x = 0$$

The new value is either 0 or 17.

Practice 2

1. The probability of selecting a spade is $\frac{3}{11}$. This probability is equal to the number of spades, 9, divided by the total number of cards. We are looking for the number of diamonds, so let x represent the number of diamonds. The total number of cards is equal to $7 + 9 + 11 + x = 27 + x$. The probability of selecting a spade is $\frac{9}{27+x}$. Set $\frac{3}{11}$ equal to $\frac{9}{27+x}$. Cross multiply and set the products equal to each other:

$$\frac{3}{11} = \frac{9}{27+x}$$
$$3(27 + x) = 99$$
$$81 + 3x = 99$$
$$3x = 18$$
$$x = 6$$

There are 6 diamonds in the deck.

2. The probability of selecting a baseball is equal to the number of baseballs divided by the total number of balls. Let x represent the number of baseballs added to the sack. There are now $4 + x$ baseballs, and $4 + x + 5 + 6 = 15 + x$ total balls. The probability of selecting a baseball is equal to $\frac{4+x}{15+x}$.

The value of x will make this probability equal to $\frac{3}{4}$, so set the two fractions equal to each other, cross multiply, and solve for x:

$$\frac{3}{4} = \frac{4+x}{15+x}$$
$$4(4+x) = 3(15+x)$$
$$16 + 4x = 45 + 3x$$
$$16 + x = 45$$
$$x = 29$$

For the probability of selecting a baseball to become $\frac{3}{4}$, 29 baseballs must be added to the sack.

3. The probability of selecting a nickel is equal to the number of nickels divided by the total number of coins. Let x represent the number of nickels removed from the jar. There are now $16 - x$ nickels, and $14 + 16 - x + 6 + 12 = 48 - x$ total coins. The probability of selecting a nickel is equal to $\frac{16-x}{48-x}$. The value of x will make this probability equal to $\frac{1}{5}$, so set the two fractions equal to each other, cross multiply, and solve for x:

$$\frac{1}{5} = \frac{16-x}{48-x}$$
$$48 - x = 5(16-x)$$
$$48 - x = 80 - 5x$$
$$48 + 4x = 80$$
$$4x = 32$$
$$x = 8$$

For the probability of selecting a nickel to become $\frac{1}{5}$, 8 nickels must be removed from the jar.

using algebra: percents and simple interest

*Mathematics is made of 50 percent formulas,
50 percent proofs, and 50 percent imagination.*

—AUTHOR UNKNOWN

In this lesson, you'll learn how to use algebra to find the percent increase or percent decrease between two values and how to calculate simple interest.

A PERCENT IS A NUMBER out of one hundred. For example, 36% is 36 out of 100. We can answer some simple percent questions without algebra:

What is 10% of 60? 10% = 0.10. Multiply 60 by 0.10: (60)(0.10) = 6. Therefore, 6 is 10% of 60.

What percent is 12 of 48? Divide 12 by 48: 12 ÷ 48 = 0.25, 0.25 = 25%. Therefore, 12 is 25% of 48.

Finally, 20% of what number is 15? 20% = 0.20. Divide 15 by 0.20: 15 ÷ 0.20 = 75. Therefore, 15 is 20% of 75.

PERCENT INCREASE

When percent questions get a bit tougher, we can use algebra to help us find the answers. If a value increases from 12 to 15, we can find the percent increase by subtracting the original value from the new value and dividing by the original value: $\frac{15-12}{12} = \frac{3}{12} = 0.25 = 25\%$.

But what if we were given the original value and the percent increase, and we needed to find the new value? We can use the same formula, but let x represent the new value.

Example

What is 20 after a 40% increase?

The original value is 20 and the new value is x. Because the percent increase is equal to the new value minus the old value divided by the old value, subtract 20 from x and divide by 20. Set that fraction equal to 40%, which is 0.40:

$$\frac{x-20}{20} = 0.40$$

Multiply both sides of the equation by 20, and then add 20 to both sides:

$$x - 20 = 8$$
$$x = 28$$

Algebra also comes in handy if we are given a new value after a percent increase and we are asked to find the original value.

Example

After a 24% increase, a value is now 80.6. What was the original value?

We can use the same formula that we used to solve the preceding question, but this time, we let x represent the original value instead of the new value. The difference between the new value and the original value is $80.6 - x$. We divide that by the original value, x, and set it equal to the percent increase, 24%, or 0.24:

$$\frac{80.6-x}{x} = 0.24$$

Multiply both sides of the equation by x, and then add x to both sides:

$$0.24x = 80.6 - x$$
$$1.24x = 80.6$$

Divide both sides by 1.24:

$$x = 65$$

PERCENT DECREASE

The formula for percent decrease is very similar to the one for percent increase. If a value decreases from 15 to 12, we can find the percent decrease by subtracting the new value from the original value and dividing by the original value: $\frac{15-12}{15} = \frac{3}{15} = 0.20 = 20\%$.

....

TIP: When finding a percent increase or decrease, be sure to divide the difference between the two values by the original value. The percent increase from one value to a second value is not equal to the percent decrease from that second value to the first value. Looking back at our two examples, going from 12 to 15 was a 25% increase, but going from 15 to 12 was only a 20% decrease. In both examples, the difference between 12 and 15 is 3, but in the percent increase example, we divided 3 by 12, because 12 was the original value, and in the percent decrease example, we divided 3 by 15, because 15 was the original value.

....

Example

After a 68% decrease, a value is now 74. What was the original value?

Let x represent the original value. The difference between the original value and the new value is $x - 74$. Divide that by the original value, x, and set it equal to the percent decrease, 68%, which is 0.68:

$$\frac{x-74}{x} = 0.68$$

Multiply both sides of the equation by x, and then subtract x from both sides:

$$0.68x = x - 74$$
$$-0.32x = -74$$

Divide both sides by –0.32:

$$x = 231.25$$

Practice 1

Answer each of the following questions.

1. What is 22 after a 50% increase?

2. What number, after a 35% increase, is 124.2?

3. What number, after a 12% increase, is 60.48?

4. What is 84 after a 75% decrease?

5. What number, after a 92% decrease, is 62?

6. What number, after a 52% decrease, is 84?

SIMPLE INTEREST

We can find how much interest an amount of money, or principal, has gained by multiplying the principal by an interest rate and a length of time. The formula for interest is $I = prt$. Interest, principal, rate, and time are all variables, and given any three of them, we can substitute those values into the equation to find the missing fourth value.

Example
If a principal of $500 gains $60 in interest in three years, what was the interest rate per year?

   The interest rate is the percent of the principal that is added as interest each year. Because $I = prt$ and we are looking for the rate, r, we can divide both sides of the equation by pt to get r alone on the right side: $\frac{I}{pt} = r$. To find the rate, divide the interest by the product of the principal and the time: $60 \div (500)(3) = 60 \div 1,500 = 0.04 = 4\%$. The principal gained interest at a rate of 4% per year.

....

TIP: When calculating interest, be sure the interest rate and the time have consistent units of measure. If the interest rate is given on a yearly basis, the time must also be in years. If the interest rate is given on a yearly basis and the time is given in months, convert the time to years before using the interest formula.

....

Example
If $36 in interest is gained over six months at a rate of 6% per year, how much was the principal?

   We can rewrite the formula $I = prt$ to solve for p by dividing both sides of the equation by rt: $\frac{I}{rt} = p$. The interest rate is given per year, but the length of time is given in months. Divide the number of months by 12, because there are 12 months in a year: $6 \div 12 = 0.5$. The time is 0.5 years and the rate is 6%, or 0.06. Because $\frac{I}{rt} = p$, $p = \frac{36}{(0.06)(0.5)} = \frac{36}{0.03} = \$1,200$. The principal was $1,200.

Practice 2

Answer each of the following questions.

1. If $2,000 gains $720 in interest at a rate of 4.5% per year, for how long was the principal in the bank?

2. If $1,800 gains $882 in interest in seven years, what was the interest rate per year?

3. If $52 in interest is gained over three months at a rate of 5.2% per year, what was the principal?

ANSWERS

Practice 1

1. The original value is 22 and the new value is x. Percent increase is equal to the new value minus the original value divided by the original value. Subtract 22 from x and divide by 22. Set that fraction equal to 50%, which is 0.50:
$$\frac{x-22}{22} = 0.50$$
Multiply both sides of the equation by 22, and then add 22 to both sides:
$$x - 22 = 11$$
$$x = 33$$
Therefore, 22 after a 50% increase is 33.
2. The original value is unknown, so let x represent the original value. The new value is 124.2. Percent increase is equal to the new value minus the original value divided by the original value.
Subtract x from 124.2 and divide by x. Set that fraction equal to 35%, which is 0.35:
$$\frac{124.2-x}{x} = 0.35$$
Multiply both sides of the equation by x, and then add x to both sides:
$$0.35x = 124.2 - x$$
$$1.35x = 124.2$$
Divide both sides by 1.35:
$$x = 92$$
Therefore, 92, after a 35% increase is 124.2.

3. The original value is unknown, so let x represent the original value. The new value is 60.48. Percent increase is equal to the new value minus the original value divided by the original value.

   Subtract x from 60.48 and divide by x. Set that fraction equal to 12%, which is 0.12:

   $$\frac{60.48 - x}{x} = 0.12$$

   Multiply both sides of the equation by x, and then add x to both sides:

   $$0.12x = 60.48 - x$$
   $$1.12x = 60.48$$

   Divide both sides by 1.12:

   $$x = 54$$

   Therefore, 54, after a 12% increase is 60.48.

4. The original value is 84 and the new value is x. Percent decrease is equal to the original value minus the new value divided by the original value.

   Subtract x from 84 and divide by 84. Set that fraction equal to 75%, which is 0.75:

   $$\frac{84 - x}{84} = 0.75$$

   Multiply both sides of the equation by 84, and then subtract 84 from both sides:

   $$84 - x = 63$$
   $$-x = -21$$
   $$x = 21$$

   Therefore, 84, after a 75% decrease, is 21.

5. The original value is unknown, so let x represent the original value. The new value is 62. Percent decrease is equal to the original value minus the new value divided by the original value.

   Subtract 62 from x and divide by x. Set that fraction equal to 92%, which is 0.92:

   $$\frac{x - 62}{x} = 0.92$$

   Multiply both sides of the equation by x, and then subtract x from both sides:

   $$0.92x = x - 62$$
   $$-0.08x = -62$$

   Divide both sides by -0.08:

   $$x = 775$$

   Therefore, 775 after a 92% decrease is 62.

6. The original value is unknown, so let x represent the original value. The new value is 84. Percent decrease is equal to the original value minus the new value divided by the original value.

Subtract 84 from x and divide by x. Set that fraction equal to 52%, which is 0.52:

$$\frac{x-84}{x} = 0.52$$

Multiply both sides of the equation by x, and then subtract x from both sides:

$$0.52x = x - 84$$
$$-0.48x = -84$$

Divide both sides by −0.48:

$$x = 175$$

Therefore, 175 after a 52% decrease is 84.

Practice 2

1. Because $I = prt$ and we are looking for the time, t, we can divide both sides of the equation by pr to get t alone on the right side: $\frac{I}{pr} = t$. The rate, 4.5%, as a decimal is 0.045.

Substitute 720 for I, 2,000 for p, and 0.045 for r:

$$t = \frac{720}{(2,000)(0.045)} = \frac{720}{90} = 8$$

The principal was in the bank for 8 years.

2. Because $I = prt$ and we are looking for the rate, r, we can divide both sides of the equation by pt to get r alone on the right side: $\frac{I}{pt} = r$.

Substitute 882 for I, 1,800 for p, and 7 for t:

$$r = \frac{882}{(1,800)(7)} = \frac{88}{12,600} = 0.07 = 7\%$$

The interest rate per year was 7%.

3. Because $I = prt$ and we are looking for the principal, p, we can divide both sides of the equation by rt to get p alone on the right side: $\frac{I}{rt} = p$.

The interest rate is given in years, but the time is given in months. There are 12 months in a year, so convert the time to years by dividing the number of months by 12: $3 \div 12 = 0.25$. The rate, 5.2%, as a decimal is 0.052.

Substitute 52 for I, 0.052 for r, and 0.25 for t:

$$p = \frac{52}{(0.052)(0.25)} = \frac{52}{0.013} = 4,000$$

The principal was $4,000.

using algebra: sequences

You can be moved to tears by numbers—
provided they are encoded and decoded fast enough.
—RICHARD DAWKINS (1941–   )
BRITISH SCIENTIST

In this lesson, you'll learn how to use algebra to find the nth term of arithmetic and geometric sequences, and how to find the exact values of terms within arithmetic and geometric sequences that contain algebraic terms.

A SEQUENCE IS A SET OF NUMBERS in which each number is generated according to a rule. We can use algebra to help us find the rule of a sequence or a certain term in a sequence. It might be easy to find the next term in a sequence once we have found the rule of the sequence, but what if we knew only a few numbers and wanted to find the hundredth term of the sequence? We wouldn't want to list all 100 terms!

ARITHMETIC SEQUENCES

An **arithmetic sequence** is a sequence in which the *difference* between any term and the term that precedes it is always the same. For instance, the sequence 3, 7, 11, 15, 19, . . . is an arithmetic sequence. We can find the rule of an arithmetic sequence by taking the difference between any two consecutive terms. In this sequence, the difference between consecutive terms is 4. The rule is +4, so the next term after 19 is 19 + 4, which is 23.

The numbers in a sequence are called terms. The first term of the sequence 3, 7, 11, 15, 19, . . . is 3, so we say that $t_1 = 3$. The second term in the sequence is represented as t_2 and the third as t_3, and we say that the nth term is t_n. The n represents an unknown place in the sequence. We use t_n to make a general statement about a sequence.

We found the sixth term, t_6, of the sequence by adding 4, the difference between any pair of consecutive terms, to the previous term, t_5. If we know the first term in an arithmetic sequence and the difference between consecutive terms, we can find any term in the sequence using this formula: $t_n = t_1 + (n-1)d$. In this formula, d represents the difference between any term and its previous term, and n represents the place of the term in the sequence.

We already know that the fifth term in the sequence is 19, but let's test the formula to be sure that it works. The first term, t_1, is 3. The difference, d, is 4. We are looking for the fifth term, $n = 5$:

$$t_n = t_1 + (n-1)d$$
$$t_5 = 3 + (5-1)4$$
$$t_5 = 3 + (4)4$$
$$t_5 = 3 + 16$$
$$t_5 = 19$$

What if the terms of a sequence are algebraic? Use two consecutive terms from the sequence and find their difference. Then, find the nth term of the sequence in terms of the variables used in the sequence.

Example
$x + 2, 3x, 4x + 3, 6x + 1, \ldots$

We can find the fifth term, sixth term, or any other term using the formula $t_n = t_1 + (n-1)d$. Let's find the seventh term.

First, find the difference between any pair of consecutive terms. Subtract the first term from the second term: $3x - (x+2) = 2x - 2$, so $d = 2x - 2$. The first term, t_1, is $x + 2$. Because we are looking for the seventh term, $n = 7$:

$$t_7 = (x+2) + (7-1)(2x-2)$$
$$t_7 = (x+2) + (6)(2x-2)$$
$$t_7 = (x+2) + (12x-12)$$
$$t_7 = 13x - 10$$

The seventh term of the sequence is $13x - 10$. We can find the exact value of that term, because we know that the difference between any two consecutive terms in an arithmetic sequence is always the same. That's why it didn't matter which pair of consecutive terms we chose to subtract in order to find d. We found that the difference between the first term and the second term was $2x - 2$. The difference between the second term and the third term is $(4x + 3x) - 3x = x + 3$. Because the difference between terms is always the same, $2x - 2$ must equal $x + 3$. Set these two differences equal to each other and solve for x:

$$2x - 2 = x + 3$$
$$x - 2 = 3$$
$$x = 5$$

The value of x in this sequence is 5, which means that the seventh term, $13x - 10$, is equal to $13(5) - 10 = 65 - 10 = 55$.

....

TIP: When finding two different expressions that represent the difference, be sure to subtract terms in the order they appear in the sequence. The difference between the first two terms is the second term minus the first term, and the difference between the third term and the second term is the third term minus the second term.

....

Practice 1

Answer each of the following questions.

1. Find the 20th term in the sequence 2, 7, 12, 17, . . .

2. Find the tenth term in the sequence 10, 6, 2, –2, . . .

3. Find the value of the eighth term in the sequence $5x$, $6x + 1$, $8x$, $10x - 1$, . . .

4. Find the value of the fifth term in the sequence $4x - 1$, $2x + 4$, $x + 2$, $x - 7$, . . .

5. Find the value of the seventh term in the sequence $10x - 1$, $9x + 1$, $7x$, $4x - 4$, . . .

GEOMETRIC SEQUENCES

A **geometric sequence** is a sequence in which the *ratio* between any term and the term that precedes it is always the same. For instance, the sequence 2, 4, 8, 16, 32, . . . is a geometric sequence. We can find the ratio, or rule, of a geometric sequence by dividing a term by the term that precedes it. In this sequence, the ratio between consecutive terms is 2. Because each term is equal to two times the previous term, the next term after 32 is 32(2), which is 64.

Just as we have a formula to find the nth term of an arithmetic sequence, we have a formula to find the nth term of a geometric sequence: $t_n = t_1(r^{n-1})$. In this formula, r represents the ratio of the sequence, and it is raised to the power $n - 1$. To find the eighth term of the sequence 2, 4, 8, 16, 32, . . . , we replace t_1 with 2, n with 8, and r with 2:

$$t_n = t_1(r^{n-1})$$
$$t_8 = 2(2^{8-1})$$
$$t_8 = 2(2^7)$$
$$t_8 = 2(128)$$
$$t_8 = 256$$

We can work with geometric sequences that have algebraic terms, too.

Example
Find the sixth term of the sequence $x - 4$, $3x$, $15x$, $75x$, . . .

....

TIP: When choosing two terms to use in order to find the ratio of a geometric sequence, try to pick two terms that, when divided, cause the variables to cancel out. If consecutive terms contain x with no constant added or subtracted, use these terms to find the ratio.

....

Begin by finding the ratio between consecutive terms. If we divide the third term by the second term, the variable x will cancel out, because $15x \div 3x = 5$. The ratio, r, is 5. The first term in the sequence, t_1, is $x - 4$. Because we are looking for the sixth term, $n = 6$:

$$t_n = t_1(r^{n-1})$$
$$t_6 = (x - 4)(5^{6-1})$$
$$t_6 = (x - 4)(5^5)$$

$$t_6 = (x - 4)(3{,}125)$$
$$t_6 = 3{,}125x - 12{,}500$$

We know that the ratio between any two terms in the sequence is 5. That means that the second term, $3x$, divided by the first term, $x - 4$, is equal to 5. Solve this equation for x:

$$\frac{3x}{x-4} = 5$$
$$(x - 4)\frac{3x}{x-4} = (x - 4)5$$
$$3x = 5x - 20$$
$$-2x = -20$$
$$x = 10$$

Because $x = 10$, the sixth term is equal to $3{,}125(10) - 12{,}500 = 31{,}250 - 12{,}500 = 18{,}750$.

It was helpful in finding the ratio that the variable x canceled out when we divided the third term by the second term. The fourth term divided by the third term is also 5. If the second term divided by the first term had been 5, we would have had no way of finding the value of x, because the ratios between the pairs of consecutive terms would have reduced to the same expression. If we are to find the value of the variable, we must always be given a sequence in which the ratio between two terms is a different algebraic expression from the ratios between the other pairs of terms. This is also true with arithmetic expressions: If the difference between every pair of consecutive terms is the same algebraic expression, then we cannot find the value of the variable.

Practice 2

Answer each of the following questions.

1. Find the seventh term in the sequence 4, 12, 36, 108, . . .

2. Find the value of the sixth term in the sequence $x - 1$, $2x$, $8x$, $30x + 4$, . . .

3. Find the value of the eighth term in the sequence $15x + 8$, $8x$, $4x$, $x + 8$, . . .

4. Find the value of the fifth term in the sequence $-\frac{1}{18}x$, $x - 2$, $-2x$, $12x$, . . .

ANSWERS

Practice 1

1. The formula $t_n = t_1 + (n-1)d$ gives us the value of any term in an arithmetic sequence. The first term, t_1, is 2.
   The difference, d, can be found by subtracting any pair of consecutive terms: $7 - 2 = 5$, so $d = 5$. Because we are looking for the 20th term, $n = 20$:
   $$t_{20} = 2 + (20 - 1)5$$
   $$t_{20} = 2 + (19)5$$
   $$t_{20} = 2 + 95$$
   $$t_{20} = 97$$
   The 20th term in the sequence is 97.

2. The formula $t_n = t_1 + (n-1)d$ gives us the value of any term in an arithmetic sequence. The first term, t_1, is 10.
   The difference, d, can be found by subtracting any pair of consecutive terms: $6 - 10 = -4$, so $d = -4$. Because we are looking for the tenth term, so $n = 10$:
   $$t_{10} = 10 + (10 - 1)(-4)$$
   $$t_{10} = 10 + (9)(-4)$$
   $$t_{10} = 10 - 3x6$$
   $$t_{10} = -26$$
   The tenth term in the sequence is -26.

3. The formula $t_n = t_1 + (n-1)d$ gives us the value of any term in an arithmetic sequence. The first term, t_1, is $5x$.
   The difference, d, can be found by subtracting any pair of consecutive terms: $(6x + 1) - 5x = x + 1$, so $d = x + 1$. Because we are looking for the eighth term, $n = 8$:
   $$t_8 = 5x + (8 - 1)(x + 1)$$
   $$t_8 = 5x + (7)(x + 1)$$
   $$t_8 = 5x + (7x + 7)$$
   $$t_8 = 12x + 7$$
   The eighth term in the sequence is $12x + 7$.
   The difference between any two consecutive terms in an arithmetic sequence is always the same. The difference between the second term and the first term is $x + 1$. The difference between the third term and the second term is $8x - (6x + 1) = 2x - 1$.

Set these two differences equal to each other and solve for x:

$$x + 1 = 2x - 1$$
$$1 = x - 1$$
$$2 = x$$

The value of x in this sequence is 2, which means that the eighth term, $12x + 7$, is equal to $12(2) + 7 = 24 + 7 = 3x1$.

4. The formula $t_n = t_1 + (n - 1)d$ gives us the value of any term in an arithmetic sequence. The first term, t_1, is $4x - 1$.

The difference, d, can be found by subtracting any pair of consecutive terms: $(2x + 4) - (4x - 1) = -2x + 5$, so $d = -2x + 5$. Because we are looking for the fifth term, $n = 5$:

$$t_5 = (4x - 1) + (5 - 1)(-2x + 5)$$
$$t_5 = (4x - 1) + (4)(-2x + 5)$$
$$t_5 = (4x - 1) - (8x + 20)$$
$$t_5 = -4x + 19$$

The fifth term in the sequence is $-4x + 19$.

The difference between any two consecutive terms in an arithmetic sequence is always the same. The difference between the second term and the first term is $-2x + 5$. The difference between the third term and the second term is $(x + 2) - (2x + 4) = -x - 2$.

Set these two differences equal to each other and solve for x:

$$-2x + 5 = -x - 2$$
$$5 = x - 2$$
$$7 = x$$

The value of x in this sequence is 7, which means that the fifth term, $-4x + 19$, is equal to $-4(7) + 19 = -28 + 19 = -9$.

5. The formula $t_n = t_1 + (n - 1)d$ gives us the value of any term in an arithmetic sequence. The first term, t_1, is $10x - 1$.

The difference, d, can be found by subtracting any pair of consecutive terms: $(9x + 1) - (10x - 1) = -x + 2$, so $d = -x + 2$. Because we are looking for the seventh term, $n = 7$:

$$t_7 = (10x - 1) + (7 - 1)(-x + 2)$$
$$t_7 = (10x - 1) + (6)(-x + 2)$$
$$t_7 = (10x - 1) - (6x + 12)$$
$$t_7 = 4x + 11$$

The seventh term in the sequence is $4x + 11$.

The difference between any two consecutive terms in an arithmetic sequence is always the same. The difference between the second term and the first term is $-x + 2$. The difference between the third term and the second term is $7x - (9x + 1) = -2x - 1$.

Set these two differences equal to each other and solve for x:

$$-x + 2 = -2x - 1$$
$$x + 2 = -1$$
$$x = -3x$$

The value of x in this sequence is $-3x$, which means that the seventh term, $4x + 11$, is equal to $4(-3x) + 11 = -12 + 11 = -1$.

Practice 2

1. The formula $t_n = t_1(r^{n-1})$ gives us the value of any term in a geometric sequence. The first term, t_1, is 4.
   The ratio, r, can be found by dividing any pair of consecutive terms: $12 \div 4 = 3x$, so $r = 3x$. Because we are looking for the seventh term, $n = 7$:
   $$t_7 = 4(3x^{7-1})$$
   $$t_7 = 4(3x^6)$$
   $$t_7 = 4(729)$$
   $$t_7 = 2{,}916$$
   The seventh term in the sequence is 2,916.

2. The formula $t_n = t_1(r^{n-1})$ gives us the value of any term in a geometric sequence. The first term, t_1, is $x - 1$.
   The ratio, r, can be found by dividing any pair of consecutive terms. Divide the third term by the second term, because this will cause the variable x to be canceled: $8x \div 2x = 4$, so $r = 4$. Because we are looking for the sixth term, $n = 6$:
   $$t_6 = (x - 1)(4^{6-1})$$
   $$t_6 = (x - 1)(4^5)$$
   $$t_6 = (x - 1)(1{,}024)$$
   $$t_6 = 1{,}024x - 1{,}024$$
   The ratio between any two consecutive terms in the sequence is 4. That means that the second term, $2x$, divided by the first term, $x - 1$, is equal to 4. Solve this equation for x:
   $$\frac{2x}{x-1} = 4$$
   $$(x - 1)\frac{2x}{x-1} = (x - 1)4$$
   $$2x = 4x - 4$$
   $$-2x = -4$$
   $$x = 2$$
   The value of x in this sequence is 2, which means that the sixth term, $1{,}024x - 1{,}024$, is equal to $1{,}024(2) - 1{,}024 = 2{,}048 - 1{,}024 = 1{,}024$.

3. The formula $t_n = t_1(r^{n-1})$ gives us the value of any term in a geometric sequence. The first term, t_1, is $15x + 8$.

The ratio, r, can be found by dividing any pair of consecutive terms. Divide the third term by the second term, because this will cause the variable x to be canceled: $4x \div 8x = \frac{1}{2}$, so $r = \frac{1}{2}$. Because we are looking for the eighth term, $n = 8$:

$$t_8 = (15x + 8)(\tfrac{1}{2^{8-1}})$$

$$t_8 = (15x + 8)(\tfrac{1}{2^7})$$

$$t_8 = (15x + 8)(\tfrac{1}{128})$$
$$t_8 = \tfrac{15x + 8}{128}$$

The ratio between any two consecutive terms in the sequence is $\frac{1}{2}$. That means that the second term, $8x$, divided by the first term, $15x + 8$, is equal to $\frac{1}{2}$. Solve this equation for x:

$$\tfrac{8x}{15x + 8} = \tfrac{1}{2}$$
$$16x = 15x + 8$$
$$x = 8$$

The value of x in this sequence is 8, which means that the eighth term, $\frac{15x + 8}{128}$, is equal to $\frac{15(8) + 8}{128} = \frac{120 + 8}{128} = \frac{128}{128} = 1$.

4. The formula $t_n = t_1(r^{n-1})$ gives us the value of any term in a geometric sequence. The first term, t_1, is $-\frac{1}{18}x$.

The ratio, r, can be found by dividing any pair of consecutive terms. Divide the fourth term by the third term, because this will cause the variable x to be canceled: $12x \div -2x = -6$, so $r = -6$. Because we are looking for the fifth term, $n = 5$:

$$t_5 = (-\tfrac{1}{18}x)(-6^{5-1})$$
$$t_5 = (-\tfrac{1}{18}x)(-6^4)$$
$$t_5 = (-\tfrac{1}{18}x)(1{,}296)$$
$$t_5 = -72x$$

The ratio between any two consecutive terms in the sequence is -6. That means that the third term, $-2x$, divided by the second term, $x - 2$, is equal to -6. Solve this equation for x:

$$\tfrac{-2x}{x - 2} = -6$$
$$-2x = -6x + 12$$
$$4x = 12$$
$$x = 3$$

The value of x in this sequence is 3, which means that the fifth term, $-72x$, is equal to $-72(3) = -216$.

using algebra: geometry

A circle is the longest distance to the same point.
—TOM STOPPARD (1937–   )
BRITISH PLAYWRIGHT

In this lesson, you'll learn how to use algebra to find the perimeter, area, and volume of two- and three-dimensional figures.

GEOMETRY IS FILLED WITH FORMULAS—perimeter of a square, area of a triangle, volume of a sphere—and any time you have a formula and an unknown value, you can use algebra, even in geometry, to find that value.

PERIMETER

The perimeter of a square is equal to 4 times the length of one side of a square: $P = 4s$. If we know the length of one side, we can substitute it for s to find P. If we know the perimeter, we can divide by 4 to find the length of one side. A square that has a perimeter of 56 inches has sides that each measure $56 \div 4 = 14$ inches.

If the length of one side of a square is an algebraic expression, we can express its perimeter by multiplying that expression by 4. A square whose sides each measure $2x$ inches has a perimeter of $4(2x) = 8x$ inches.

Example

A square has a perimeter of 100 yards. If the length of one side is equal to $(6x + 1)$ yards, what is the value of x?

We use the formula $P = 4s$, and replace P with 100 and s with $6x + 1$:

$$100 = 4(6x + 1)$$
$$100 = 24x + 4$$
$$96 = 24x$$
$$x = 4$$

TIP: If you are missing two values in a formula, let x equal one of those values and try to write the other value in terms of x, so that only one variable is used. Remember, you cannot find the value of two variables if you have only one equation.

The perimeter of a rectangle is equal to twice its length plus twice its width: $P = 2l + 2w$. The length and width of a rectangle might be given to us as algebraic expressions.

Example

The length of a rectangle is 5 centimeters more than 3 times its width. If the perimeter of the rectangle is 82 centimeters, what is the length of the rectangle?

We don't know the value of either the width or the length of the rectangle. But, if we let x represent the width, we can represent the length as $3x + 5$, because the length is 5 more than 3 times the width. Now, we can substitute these values into the formula for the perimeter of a rectangle:

$$P = 2l + 2w$$
$$82 = 2(3x + 5) + 2x$$
$$82 = 6x + 10 + 2x$$
$$82 = 8x + 10$$
$$72 = 8x$$
$$9 = x$$

Because x represents the width of the rectangle, the width of the rectangle is 9 centimeters. The length is 5 centimeters more than 3 times the width: $3(9) + 5 = 27 + 5 = 32$ centimeters.

AREA AND VOLUME

We can perform the same sort of substitutions for area and volume formulas to find the length of a side or an edge of a solid. The following are some common area and volume formulas:

$A_{square} = s^2$, where s is the length of one side of the square

$A_{rectangle} = lw$, where l is the length and w is the width of the rectangle

$A_{triangle} = \frac{1}{2}bh$, where b is the base and h is the height of the triangle

$A_{circle} = \pi r^2$, where r is the radius of the circle

$V_{cube} = e^3$, where e is the length of one edge of the cube

$V_{rectangular\ prism} = lwh$, where l is the length, w is the width, and h is the height of the prism

$V_{cylinder} = \pi r^2 h$, where r is the radius and h is the height of the cylinder

$V_{cone} = \frac{1}{3}\pi r^2 h$, where r is the radius and h is the height of the cone

$V_{sphere} = \frac{4}{3}\pi r^3$, where r is the radius of the sphere

Example

If the length of one side of a square is $(x + 8)$ units, what is the area of the square in terms of x?

The formula for area of a square is $A_{square} = s^2$, so we must square the length of one side of the square: $(x + 8)^2 = (x + 8)(x + 8)$. Use FOIL to find the area of the square:

First: $(x)(x) = x^2$
Outside: $(x)(8) = 8x$
Inside: $(8)(x) = 8x$
Last: $(8)(8) = 64$
$x^2 + 8x + 8x + 64 = x^2 + 16x + 64$.

The area of the square is $x^2 + 16x + 64$ square units.

Example

The height of a cone is twice its radius. If the volume of the cone is 18π cm^3, what is the height of the cone?

Let x equal the radius of the cone. Because the height of the cone is twice the radius, the height is $2x$. The formula for volume of a cone is $V_{cone} = \frac{1}{3}\pi r^2 h$, so substitute 18π for V, x for r, and $2x$ for h:

$$18\pi = \frac{1}{3}\pi x^2(2x)$$
$$18 = \frac{2}{3}x^3$$
$$27 = x^3$$
$$3 = x$$

The radius is 3 centimeters. Because the height is twice the radius, the height of the cone is 6 centimeters.

Practice

Answer each of the following questions.

1. The perimeter of a square is $(10x + 2)$ feet. If one side of the square is $(x + 5)$ feet, what is the value of the perimeter of the square?

2. The perimeter of a rectangle is 26 centimeters. If the length of the rectangle is 7 less than 4 times the width of the rectangle, what is the width of the rectangle?

3. Find the area of a rectangle that has a length of $(2x + 1)$ units and a width of $(3x - 4)$ units in terms of x.

4. The base and height of a triangle are two consecutive even numbers. If the area of the triangle is 40 cm^2, and the base is greater than the height, what is the base of the triangle?

5. If a circle has an area of $(x^2 + 10x + 25)\pi$ square feet, what is the radius of the circle?

6. A cylinder has a volume of 729π millimeters^3. If the radius and the height are the same, what is the measurement of each?

7. The length of a rectangular prism is half its width, which is half its height. If the volume of the prism is 512 cubic meters, what is the height of the prism?

ANSWERS

Practice

1. The formula for perimeter of a square is $P = 4s$. The perimeter of the square is $(10x + 2)$ and one side of the square is $(x + 5)$ feet, so substitute $(10x + 2)$ for P and substitute $(x + 5)$ for s:

$$(10x + 2) = 4(x + 5)$$
$$10x + 2 = 4x + 20$$
$$6x + 2 = 20$$
$$6x = 18$$
$$x = 3$$

Because $x = 3$, the perimeter of the square is $10(3) + 2 = 30 + 2 = 32$ feet.

2. The formula for perimeter of a rectangle is $P = 2l + 2w$. Let x represent the width of the rectangle. The length is 7 less than 4 times the width, which means that it is $4x - 7$. Substitute 26 for P, $4x - 7$ for l, and x for w:

$$26 = 2(4x - 7) + 2(x)$$
$$26 = 8x - 14 + 2x$$
$$26 = 10x - 14$$
$$40 = 10x$$
$$4 = x$$

Because $x = 4$, the width of the rectangle is 4 centimeters.

3. The formula for area of a rectangle is $A = lw$. Multiply the length, $2x + 1$, by the width, $3x - 4$. Use FOIL:

First: $(2x)(3x) = 6x^2$

Outside: $(2x)(-4) = -8x$

Inside: $(1)(3x) = 3x$

Last: $(1)(-4) = -4$

$6x^2 - 8x + 3x - 4 = 6x^2 - 5x - 4$ square units.

4. The formula for area of a triangle is $A = \frac{1}{2}bh$. The base and the height are two consecutive even numbers. If one of them is x, then the other must be $x + 2$, because the next even number is exactly 2 greater than x. The base is greater than the height, so substitute $x + 2$ for b and x for h. Substitute 40 for A:

$$40 = \frac{1}{2}(x + 2)(x)$$
$$40 = \frac{1}{2}(x^2 + 2x)$$
$$80 = x^2 + 2x$$
$$x^2 + 2x - 80 = 0$$

Factor this quadratic equation and find the solutions for x. The only factors of -80 that have a difference of 2 are -8 and 10:

$$(x - 8)(x + 10) = 0$$
$$x = 8, -10$$

The height cannot be a negative value, so the height of the triangle must be 8 centimeters. The next consecutive even number is 10, so the base of the triangle is 10 centimeters.

5. The formula for area of a circle is $A = \pi r^2$. Substitute $(x^2 + 10x + 25)\,\pi$ for A:

$$(x^2 + 10x + 25)\,\pi = \pi r^2$$
$$x^2 + 10x + 25 = r^2$$

Factor $x^2 + 10x + 25$. The only factors of 25 that multiply to 25 and add to 10 are 5 and 5:

$$(x + 5)(x + 5) = r^2$$
$$(x + 5)^2 = r^2$$
$$x + 5 = r$$

The radius of the circle is $(x + 5)$ feet.

6. The formula for volume of a cylinder is $V = \pi r^2 h$. The radius and the height are the same, so let x represent both of them. Substitute 729π for V:

$$729\pi = \pi\,(x^2)(x)$$
$$729 = x^3$$
$$9 = x$$

The radius and the height of the cylinder are each 9 millimeters.

7. The formula for volume of a rectangular prism is $V = lwh$. The length of the prism is half its width, which is half its height, so if x represents the length, then $2x$ is the width and $4x$ is the height. Substitute 512 for V, x for l, $2x$ for w, and $4x$ for h:

$$512 = (x)(2x)(4x)$$
$$512 = 8x^3$$
$$64 = x^3$$
$$4 = x$$

The length of the prism is 4 meters, which means that the width is $2(4) = 8$ meters and the height is $2(8) = 16$ meters.

POSTTEST

THE 30 LESSONS you have just read have given you all the skills necessary to ace the posttest. The posttest has 30 questions, and the order of these questions corresponds with the order of the lessons in this book. The questions are very similar to those in the pretest, so you can see your progress and improvement since taking the pretest.

After completing the posttest, check your answers and compare your score, question by question, to your pretest. Where did you improve, and where do you need some extra practice? For example, if you got question 15 wrong on the pretest, did you get it right on the posttest? The posttest can help you identify which areas you have mastered and which areas you need to review a bit more. Return to the lessons that cover the topics that are tough for you until those topics become your strengths.

POSTTEST

1. $20n^3 - 17n^3 =$
   a. 3
   b. $3n$
   c. $3n^2$
   d. $3n^3$

2. $(8r^3s^2t^7)(12s^4t^5) =$
   a. $96s^6t^{12}$
   b. $96r^3s^{-2}t^2$
   c. $96r^3s^4t^5$
   d. $96r^3s^6t^{12}$

3. What is $12u^2$ when $u = 3$?
   a. 9
   b. 36
   c. 108
   d. 1,296

4. Find the value of $-5j + 6(j - 4k) - 8k$ when $j = 2$ and $k = -1$.
   a. 34
   b. 30
   c. 22
   d. -65

5. Which of the following is half the difference between a number and five?
   a. $x - 5$
   b. $\frac{1}{2}(x - 5)$
   c. $\frac{1}{2}x - 5$
   d. $5 - \frac{1}{2}x$

6. Which of the following is a correct factoring of $6a^8b^2 + 9a^5bc^5$?
   a. $3a^5b(2a^3b + 3c^5)$
   b. $3a^3b(2a^5b^2 + 3a^5c^5)$
   c. $3abc(2a^7b + 3a^4c^4)$
   d. $6a^8b^2(1 + 9a^5bc^5)$

7. $6(e^{-2})^{-2} =$

   a. $6e^{-4}$

   b. $6e^4$

   c. $36e^{-4}$

   d. $36e^4$

8. If $9u = 63$, what is the value of u?

   a. -7

   b. 5

   c. 7

   d. 54

9. If $f - 8 = g + 12$, what is f in terms of g?

   a. $g + 4$

   b. $g - 8$

   c. $8g + 12$

   d. $g + 20$

10. What is the value of v in $-2(3v + 5) = 14$?

   a. -4

   b. -2

   c. 1

   d. 3

11. In the equation $\frac{3c}{4} - 9 = 3$, what is the value of c?

   a. 4

   b. 12

   c. 16

   d. 20

12. What is the value of w in $3\sqrt{12w} = 18$?

   a. 2

   b. 3

   c. 6

   d. 36

13. Which of the following lines is parallel to the line $y = -\frac{1}{8}x + 1$?

    a. $y = -\frac{1}{8}x$

    b. $y = \frac{1}{8}x + 5$

    c. $y = 8x + 4$

    d. $y = -8x - 1$

14. What value completes the following table?

x	y
−4	15
−2	11
2	?
5	−3
7	−7

    a. −11

    b. 0

    c. 3

    d. 8

15. Which of the following is the graph of $y = -3$?

    a.

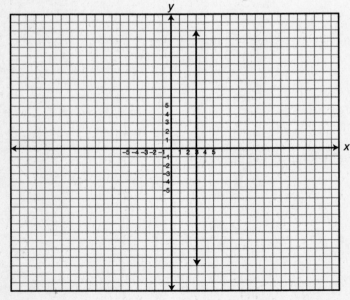

b.

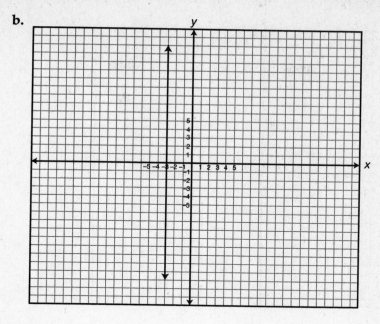

c.

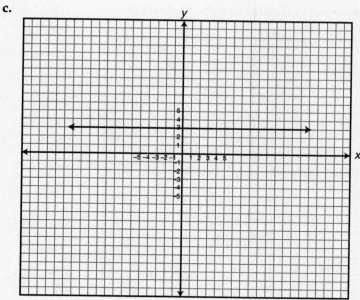

d.

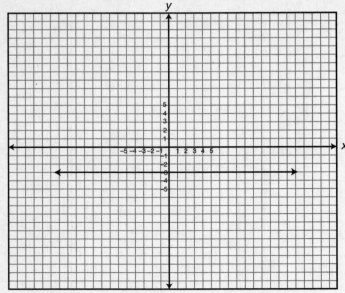

16. What is the equation of the line graphed below?

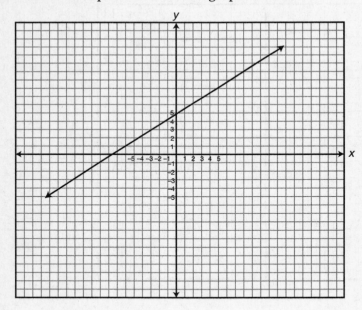

a. $y = -\frac{2}{3}x + 5$

b. $y = \frac{2}{3}x + 5$

c. $y = -\frac{3}{2}x + 5$

d. $y = \frac{3}{2}x + 5$

17. What is the distance from (−5,4) to (4,−7)?
   a. $\sqrt{10}$ units
   b. $3\sqrt{10}$ units
   c. $\sqrt{122}$ units
   d. $\sqrt{202}$ units

18. What is the domain of the function $y = \frac{-1}{x+8}$?
   a. all real numbers
   b. all real numbers except −1
   c. all real numbers except −8
   d. all real numbers except 8

19. What is the solution to the following system of equations?
$$5y - 2x = 1$$
$$7y + 2x = -13$$
   a. $x = 7, y = 3$
   b. $x = -3, y = -1$
   c. $x = 4, y = -3$
   d. $x = -2, y = 1$

20. What is the solution set of $-8x + 11 < 83$?
   a. $x > -9$
   b. $x < -9$
   c. $x > 9$
   d. $x < 9$

21. What inequality is shown on the following graph?

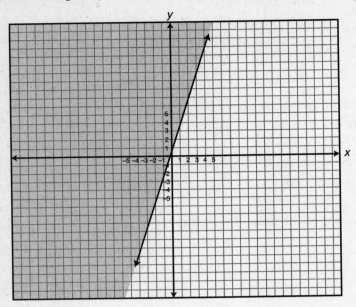

   a. $y \geq 3x + 1$
   b. $y > 3x + 1$
   c. $y < 3x + 1$
   d. $y \leq 3x + 1$

22. $(10x + 4)(5x - 2) =$
   a. $50x^2 - 8$
   b. $50x^2 - 20x - 8$
   c. $50x^2 + 20x - 8$
   d. $50x^2 + 40x - 8$

23. $25x^{16} - 144$ is the product of which factors?
   a. $(5x^4 + 12)(5x^4 - 12)$
   b. $(5x^8 + 12)(5x^8 - 12)$
   c. $(5x^8 + 12)(5x^8 + 12)$
   d. $(5x^8 - 12)(5x^8 - 12)$

24. What are the roots of $x^2 + 7x + 5 = 0$?

   a. $x = -5, x = -1$

   b. $x = 5, x = 1$

   c. $x = -7 + \sqrt{29}, x = -7 - \sqrt{29}$

   d. $x = \dfrac{-7 + \sqrt{29}}{2}, x = \dfrac{-7 - \sqrt{29}}{2}$

25. Sherry's farm is 6 acres. She buys one-eighth of Nancy's farm, increasing the size of her farm to 10 acres. What was the size of Nancy's farm?

   a. 4 acres

   b. 16 acres

   c. 32 acres

   d. 128 acres

26. The ratio of students who wear contact lenses to students who wear glasses at Hunter Middle School is 7:2. If the number of students who wear contact lenses is 36 less than four times the number of students who wear glasses, how many students wear glasses?

   a. 36

   b. 72

   c. 144

   d. 252

27. Kara has 10 cans of tomato soup, 8 cans of chicken noodle soup, and 4 cans of mushroom soup, and she will choose one can of soup at random for lunch. How many cans of mushroom soup must she buy to make the probability of selecting a can of mushroom soup $\frac{1}{4}$?

   a. 2

   b. 3

   c. 5

   d. 6

28. What number, after a 22.5% decrease, is 93?

   a. 27

   b. 114

   c. 115

   d. 120

29. Find the value of the sixth term of the following geometric sequence:

$x - 8, 2x - 6, -3x, 9x, \ldots$

   a. 53

   b. 61

   c. 486

   d. 75

30. The area of a triangle is 45 square inches. If the height is two inches less than four times the base, what is the height of the triangle?

   a. 5 inches

   b. 9 inches

   c. 18 inches

   d. 36 inches

ANSWERS

1. d. Each term has a base of n and an exponent of 3, so the base and exponent of your answer will be n^3. Subtract the coefficient of the second term from the coefficient of the first term: $20 - 17 = 3$, so $20n^3 - 17n^3 = 3n^3$. For more on this skill, review Lesson 2.

2. d. Multiply the coefficients: $(8)(12) = 96$. The first term has bases of r, s, and t and the second term has bases of s and t, so your answer will have bases of r, s, and t.

The exponent of r in the answer is 3, because there is no r in the second term. Add the exponent of s in the first term to the exponent of s in the second term: $2 + 4 = 6$. The exponent of s in the answer is 6.

Add the exponent of t in the first term to the exponent of t in the second term: $7 + 5 = 12$. The exponent of t in the answer is 12.

$$(8r^3s^2t^7)(12s^4t^5) = 96r^3s^6t^{12}.$$

For more on this skill, review Lesson 3.

3. c. Replace u with 3: $12(3)^2$.

Exponents come before multiplication, so handle the exponent first: $3^2 = 9$.

The expression becomes $12(9)$.

Multiply: $12(9) = 108$.

For more on this skill, review Lesson 4.

4. a. Use the distributive law to find $6(j - 4k)$. Multiply each term in parentheses by 6:

$$6(j - 4k) = 6j - 24k$$

This expression, $-5j + 6j - 24k - 8k$, has two j terms and one k term.

Combine the j terms: $-5j + 6j = j$.

Combine the k terms: $-24k - 8k = -32k$.

The expression is now $j - 32k$.

Substitute 2 for j and -1 for k: $2 - 32(-1) = 2 + 32 = 34$.

For more on this skill, review Lesson 5.

5. b. The keyword *half* means multiplication by $\frac{1}{2}$ (or division by 2) and the keyword *difference* signals subtraction. *The difference between a number and five* is $x - 5$. Half of that quantity is $\frac{1}{2}(x - 5)$.

For more on this skill, review Lesson 6.

6. a. $1, 2, 3, 6, a^8$, and b^2 are factors of $6a^8b^2$ and $1, 3, 9, a^5, b$, and c^5 are factors of $9a^5bc^5$.

The greatest common factor of 6 and 9 is 3. The variables a and b are common to both terms, but the variable c is not in the first term. The smaller exponent of a is 5 and the smaller exponent of b is 1, so $3a^5b$ can be factored out of every term.

Divide both terms by $3a^5b$: $6a^8b^2 \div 3a^5b = 2a^3b$, and $9a^5bc^5 \div 3a^5b = 3c^5$.

$6a^8b^2 + 9a^5bc^5$ factors into $3a^5b(2a^3b + 3c^5)$.

For more on this skill, review Lesson 7.

7. b. (e^{-2}) is raised to the negative second power, and then multiplied by 6.

Multiply -2 by -2: $(-2)(-2) = 4$, so $(e^{-2})^{-2} = e^4$.

Multiply e^4 by its coefficient, 6: $(6)(e^4) = 6e^4$.

For more on this skill, review Lesson 8.

8. c. In the equation $9u = 63$, u is multiplied by 9. Use the opposite operation, division, to solve the equation. Divide both sides of the equation by 9:

$$\frac{9u}{9} = \frac{63}{9}$$
$$u = 7$$

For more on this skill, review Lesson 9.

9. d. To find f in terms of g, we must get f alone on one side of the equation, with g on the other side of the equal sign. In the equation $f - 8 = g + 12$, 8 is subtracted from f. Use the opposite operation, addition, to get f alone on one side of the equation. Add 8 to both sides of the equation:

$$f - 8 + 8 = g + 12 + 8$$
$$f = g + 20$$

The value of f, in terms of g, is $g + 20$.

For more on this skill, review Lesson 9.

10. a. The equation $-2(3v + 5) = 14$ has a constant multiplying an expression. The first step to solving this equation is to use the distributive law.

Multiply $3v$ and 5 by -2: $-2(3v + 5) = -6v - 10$.

The equation is now: $-6v - 10 = 14$.

The equation shows subtraction and multiplication, so we must use addition and division to solve it. Add 10 to both sides of the equation:

$$-6v - 10 + 10 = 14 + 10$$
$$-6v = 24$$

Because v is multiplied by -6, divide both sides of the equation by -6:

$$\frac{-6v}{-6} = \frac{24}{-6}$$
$$v = -4$$

For more on this skill, review Lesson 10.

11. c. The equation $\frac{3c}{4} - 9 = 3$ shows multiplication and subtraction. We will need to use their opposites, division and addition, to find the value of c. We add first, because addition comes before division in the inverse of the order of operations:

$$\frac{3c}{4} - 9 + 9 = 3 + 9$$
$$\frac{3c}{4} = 12$$

Divide both sides of the equation by the coefficient $\frac{3}{4}$, which is the same as multiplying by its reciprocal $\frac{4}{3}$, to get c alone on the left side:

$$\left(\frac{4}{3}\right)\frac{3c}{4} = 12\left(\frac{4}{3}\right)$$
$$c = 16$$

For more on this skill, review Lesson 10.

12. b. In the equation $3\sqrt{12w} = 18$, three times the square root of $12w$ is equal to 18. First, divide both sides of the equation by 3:

$$\frac{3\sqrt{12w}}{3} = \frac{18}{3}$$
$$\sqrt{12w} = 6$$

To remove the radical symbol from the left side of the equation, we must raise both sides of the equation to the second power.

$$(\sqrt{12w})^2 = (6)^2$$
$$12w = 36$$

Because w is multiplied by 12, divide both sides of the equation by 12:

$$\frac{12w}{12} = \frac{36}{12}$$
$$w = 3$$

For more on this skill, review Lesson 11.

13. a. The slope of the line $y = -\frac{1}{8}x + 1$ is $-\frac{1}{8}$, because the line is in slope-intercept form, and the coefficient of x is $-\frac{1}{8}$. Any line with a slope of $-\frac{1}{8}$, including $y = -\frac{1}{8}x$, is parallel to the line $y = -\frac{1}{8}x + 1$.

For more on this skill, review Lesson 12.

14. c. First, find the equation that was used to build the table. Use the first two rows of the table to find the slope: $\frac{11-15}{-2-(-4)} = \frac{-4}{-2+4} = \frac{-4}{2} = -2$. Use the second row of the table and the equation $y = mx + b$ to find the y-intercept of the equation:

$$11 = -2(-2) + b$$
$$11 = 4 + b$$
$$b = 7$$

The equation for this table is $y = -2x + 7$. To find the value of y when $x = 2$, substitute 2 for x in the equation and solve for y:

$$y = -2(2) + 7$$
$$y = -4 + 7$$
$$y = 3$$

The missing value in the table is 3.
For more on this skill, review Lesson 13.

15. d. The graph of $y = -3$ is a horizontal line along which every point has a y value of -3. Only the graphs in choices **c** and **d** are horizontal lines, and only along the graph in choice **d** does every point have a y value of -3.
For more on this skill, review Lesson 14.

16. b. To find the equation of the line, begin by finding the slope using any two points on the line. When $x = 3$, $y = 7$, and when $x = 0$, $y = 5$. Use these points to find the slope: $\frac{5-7}{0-3} = \frac{-2}{-3} = \frac{2}{3}$. The y-intercept can be found right on the graph. The line crosses the y-axis where $y = 5$, which means that 5 is the y-intercept. This is the graph of the equation $y = \frac{2}{3}x + 5$.
For more on this skill, review Lesson 15.

17. d. Use the distance formula to find the distance between $(-5,4)$ and $(4,-7)$. Because $(-5,4)$ is the first point, x_1 will be -5 and y_1 will be 4. $(4,-7)$ is the second point, so x^2 will be 4 and y_2 will be -7:

$$D = \sqrt{(x_2 - x_1)^2 + (y_2 - y_1)^2}$$
$$D = \sqrt{(4-(-5))^2 + (-7-4)^2}$$
$$D = \sqrt{(9)^2 + (-11)^2}$$
$$D = \sqrt{81 + 121}$$
$$D = \sqrt{202} \text{ units}$$

For more on this skill, review Lesson 16.

18. c. The denominator of a fraction cannot be equal to zero, so we can substitute any real number for x in the equation $y = \frac{-1}{x+8}$ except -8, because -8 would make the fraction undefined. The domain of the equation is all real numbers except -8.
For more on this skill, review Lesson 17.

19. b. This system could be solved using either substitution or elimination. Because adding the two equations would eliminate the x terms, use elimination to solve.

Add the equations:

$$\begin{array}{r} 5y - 2x = 1 \\ + \; 7y + 2x = -13 \\ \hline 12y = -12 \end{array}$$

Divide by 12 to solve for y:

$$12y = -12$$
$$y = -1$$

Substitute -1 for y in either equation and solve for x:

$$5(-1) - 2x = 1$$
$$-5 - 2x = 1$$
$$-2x = 6$$
$$x = -3$$

The solution to this system of equations is $x = -3$, $y = -1$.

For more on this skill, review Lesson 18 and Lesson 19.

20. a. To solve the inequality, we must get x alone on one side of the inequality. Subtract 11 from both sides of the inequality and divide by -8. Because we are dividing by a negative number, change the inequality symbol from less than to greater than:

$$-8x + 11 < 83$$
$$-8x < 72$$
$$x > -9$$

For more on this skill, review Lesson 20.

21. a. The graphed line, $y = 3x + 1$, is solid, which means that the points on the line are part of the solution set. The answer must be either choice **a** or choice **d**. The point $(0,4)$ is in the solution set, so it can be used to test each inequality. It is true that $4 \geq 3(0) + 1$, because $4 \geq 1$, so this is the graph of the inequality $y \geq 3x + 1$. The same point, $(0,4)$, fails when inserted into the inequality $y \leq 3x + 1$.

For more on this skill, review Lesson 21.

22. a. To find the product of two binomials, use FOIL and combine like terms:

First: $(10x)(5x) = 50x^2$

Outside: $(10x)(-2) = -20x$

Inside: $(4)(5x) = 20x$

Last: $(4)(-2) = -8$

$$50x^2 - 20x + 20x - 8 = 50x^2 - 8$$

For more on this skill, review Lesson 22.

23. b. The binomial $25x^{16} - 144$ is the difference between two perfect squares. The coefficient of x, 25, is a perfect square, and the exponent of x is even. The constant, 144, is also a perfect square. The square root of $25x^{16}$ is $5x^8$, and the square root of 144 is 12. The square root of the first term, $5x^8$, plus the square root of the second term, 12, is the first factor: $5x^8 + 12$. The square root of the first term minus the square root of the second term is the second factor: $5x^8 - 12$. Therefore, $25x^{16} - 144$ factors to $(5x^8 + 12)(5x^8 - 12)$.

For more on this skill, review Lesson 23.

24. d. The equation $x^2 + 7x + 5 = 0$ is in the form $ax^2 + bx + c = 0$, but this trinomial cannot be factored easily because its roots are not integers. Use the quadratic formula. The coefficient of x^2 is 1, so $a = 1$. The coefficient of x is 7, so $b = 7$, and the constant is 5, so $c = 5$. Substitute these values into the quadratic formula:

$$x = \frac{-7 \pm \sqrt{(7)^2 - 4(1)(5)}}{2(1)}$$

Next, $(7)^2 = 49$ and $(4)(1)(5) = 20$:

$$x = \frac{-7 \pm \sqrt{49 - 20}}{2(1)}$$

Under the square root symbol, $49 - 20 = 29$, and in the denominator, $2(1) = 2$:

$$x = \frac{-7 \pm \sqrt{29}}{2}$$

The square root of 29 cannot be simplified. One root is equal to $\frac{-7 + \sqrt{29}}{2}$, and the other root is equal to $\frac{-7 - \sqrt{29}}{2}$.

For more on this skill, review Lesson 24.

25. c. We're looking for the original size of Nancy's farm, so we can use x to represent that number. Sherry buys one-eighth of Nancy's farm, which means that she buys $\frac{x}{8}$ acres. Sherry already had 6 acres, which means that the size of her farm is now $\frac{x}{8} + 6$. We know that her farm is 10 acres, so we can set $\frac{x}{8} + 6$ equal to 10 and solve for x:

$$\frac{x}{8} + 6 = 10$$
$$\frac{x}{8} = 4$$
$$x = 32$$

Nancy's farm was 32 acres.

For more on this skill, review Lesson 25.

26. b. The number of students who wear glasses is unknown, so represent that number with x.

The number of students who wear contact lenses is 36 less than four times the number of students who wear glasses, which means that the number of students who wear contact lenses is equal to $4x - 36$. The ratio of students who wear contact lenses to students who wear glasses is 7:2, or $\frac{7}{2}$. The ratio of actual students who wear contact lenses to actual students who wear glasses is $(4x - 36): x$, or $\frac{4x - 36}{x}$. These ratios are equal. Write a proportion and solve for x: $\frac{7}{2} = \frac{4x - 36}{x}$.

Multiply both sides by $2x$:

$$7x = 8x - 72.$$

Subtract $8x$ from both sides: $-x = -72$, so $x = 72$.

Therefore, 72 students wear glasses.

For more on this skill, review Lesson 26.

27. a. The probability of selecting a can of mushroom soup is equal to the number of cans of mushroom soup divided by the total number of cans of soup.

Let x represent the number of cans of mushroom soup Kara buys. There are now $4 + x$ cans of mushroom soup, and $4 + x + 10 + 8 = 22 + x$ total cans. The probability of selecting a can of mushroom soup is equal to $\frac{4 + x}{22 + x}$.

To find the value of x that will make this probability equal to $\frac{1}{4}$, set the two fractions equal to each other, cross multiply, and solve for x:

$$\frac{1}{4} = \frac{4 + x}{22 + x}$$
$$4(4 + x) = 1(22 + x)$$
$$16 + 4x = 22 + x$$
$$16 + 3x = 22$$
$$3x = 6$$
$$x = 2$$

For the probability of selecting a can of mushroom soup to become $\frac{1}{4}$, Kara must buy 2 cans of mushroom soup.

For more on this skill, review Lesson 27.

28. d. The original value is unknown, so let x represent the original value. The new value is 93. Percent decrease is equal to the original value minus the new value, all divided by the original value.

Subtract 93 from x and divide by x. Set that fraction equal to 22.5%, which is 0.225:

$$\frac{x-93}{x} = 0.225$$

Multiply both sides of the equation by x, and then subtract x from both sides:

$$0.225x = x - 93$$
$$-0.775x = -93$$

Divide by -0.775: $x = 120$.

Therefore, 120, after a 22.5% decrease, is 93.

For more on this skill, review Lesson 28.

29. c. The formula $t_n = t_1(r^{n-1})$ gives us the value of any term in a geometric sequence. The first term, t_1, is $x - 8$.

The ratio, r, can be found by dividing any term by the previous term. Divide the fourth term by the third term, because this will cause the variable x to be canceled: $9x \div -3x = -3$, so $r = -3$. Because we are looking for the sixth term, $n = 6$:

$$t_6 = (x - 8)(-3^{6-1})$$
$$t_6 = (x - 8)(-3^5)$$
$$t_6 = (x - 8)(-243)$$
$$t_6 = -243x + 1,944$$

The ratio of any term in the sequence to the previous term is -3. That means that the second term, $2x - 6$, divided by the first term, $x - 8$, is equal to -3. Solve this equation for x:

$$\frac{2x-6}{x-8} = -3$$
$$-3x + 24 = 2x - 6$$
$$-5x + 24 = -6$$
$$-5x = -30$$
$$x = 6$$

The value of x in this sequence is 6, which means that the sixth term, $-243x + 1,944$, is equal to $-243(6) + 1,944 = -1,458 + 1,944 = 486$.

For more on this skill, review Lesson 29.

30. c. The formula for area of a triangle is $A = \frac{1}{2}bh$. Let x represent the base of the triangle. Because the height is two less than four times the base, the height of the triangle is $4x - 2$. Substitute x for b and $4x - 2$ for h. Substitute 45 for A:

$$45 = \tfrac{1}{2}(x)(4x - 2)$$
$$45 = \tfrac{1}{2}(4x^2 - 2x)$$
$$45 = 2x^2 - x$$
$$2x2 - x - 45 = 0$$

Factor this quadratic equation and find the solutions for x.

The factors of $2x^2$ are -2, -1, 1, 2, $-x$, and x. The first terms of the binomials may be $2x$ and x.

The factors of -45 are -45, -15, -9, -5, -3, -1, 1, 3, 5, 9, 15, and 45.

We need two factors that multiply to -45, which means that we need one positive number and one negative number:

$(2x - 5)(x + 9) = 2x^2 + 13x - 45$. The middle term is too large.

$(2x - 9)(x + 5) = 2x^2 + x - 45$. The middle term is the wrong sign.

$(2x + 9)(x - 5) = 2x^2 - x - 45$.

Set each factor equal to zero and solve for x:

$$2x + 9 = 0$$
$$2x = -9$$
$$x = -4.5$$

A base cannot be a negative number, so this cannot be the answer.

$$x - 5 = 0$$
$$x = 5$$

The base of the triangle is 5 inches. Since the height is two inches less than four times the base, the height of the triangle is $4(5) - 2 = 20 - 2 = 18$ inches.

For more on this skill, review Lesson 30.

hints for taking standardized tests

THE TERM *standardized test* has the ability to produce fear in test takers. These tests are often given by a state board of education or a nationally recognized education group. Usually these tests are taken in the hope of getting accepted—whether it's for a special program, the next grade in school, or even to a college or university. Here's the good news: Standardized tests are more familiar to you than you know. In most cases, these tests look very similar to tests that your teachers may have given in the classroom. For most math standardized tests, you may come across two types of questions: multiple-choice and free-response questions. There are some practical ways to tackle both types!

TECHNIQUES FOR MULTIPLE-CHOICE QUESTIONS

The Process of Elimination

For some standardized tests, there is no guessing penalty. What this means is that you shouldn't be afraid to guess. For a multiple-choice question with four answer choices, you have a one in four chance of guessing correctly. And your chances improve if you can eliminate a choice or two.

By using the process of elimination, you will cross out incorrect answer choices and improve your odds of finding the correct answer. In order for the process of elimination to work, you must keep track of what choices you are crossing out. Cross out incorrect choices on the test booklet itself. If you don't cross out an incorrect answer, you may still think it is a possible answer. Crossing out any incorrect answers makes it easier to identify the right answer; there will be fewer places where it can hide!

Don't Supersize

Some multiple-choice questions are long word problems. You may get easily confused if you try to solve the word problems at once. Take bite-sized pieces. Read each sentence one at a time. As soon as you can solve one piece of the problem, go ahead and solve it. Then add on the next piece of information and solve this. Keep adding the information, until you have the final answer. For example:

> Joyce gets $5 for an allowance every day from Monday to Friday. She gets $8 every Saturday and Sunday. If she saves all her allowance for six weeks to buy a new bike, how much will she have?
> a. $150
> b. $240
> c. $246
> d. $254

Take bite-sized pieces of this problem. If Joyce makes $5 every day during the week, she makes $25 a week. If she makes $8 each weekend day, then that is another $16. Joyce makes a total of $41 a week.

Look at the final sentence. If she saves all her allowance for six weeks, you need to calculate 6 times 41, which is $246. There's your answer—choice c!

Use the Answer Choices as Your Tools

You are usually given four choices, one of which is correct. So, if you get stuck, try using the answer choices to jump-start your brain.

Instead of setting up an equation, plug an answer from the answer choices into the problem and see if it works. If it doesn't work, cross out that choice and move on to the next. You can often find the correct answer by trying just one or two of the answer choices!

TECHNIQUES FOR FREE-RESPONSE QUESTIONS

Show Your Work!

Make sure you show all your work. This is good for two reasons. First, some tests that use free-response math questions give you partial credit, even if your final answer is incorrect. If scorers see that you were using the right process to find an answer, you may get some credit. Second, by showing all your work, it is easier for you to review your answer by tracing back all the steps. This can help you cross out any careless calculations or silly mistakes.

TECHNIQUES FOR ALL QUESTIONS

Get Out of Your Head

Use any space in your test booklet to do your math work. When you attempt to do math in your head—even simple arithmetic—you run the chance of making a careless error. Accuracy is more important than speed, so always do your work on paper.

Understand the Question

You need to really get what a question is asking for. Let's say a problem asks you to find the value of $x + 2$. If you don't read carefully, you may solve for x correctly and assume the value of x is your answer. By not understanding what the question was asking for, you have picked the wrong answer.

Skipping Around

You may come across a question that you're not sure how to answer. In these cases, it's okay to skip the question and come back to it later. If your standardized test is timed, you don't want to waste too much time with a troublesome problem. The time might be over before you can get to questions that you would normally whiz through.

If you hit a tricky question, fold down the corner of the test booklet as a reminder to come back to that page later. When you go back to that question with a fresh eye, you may have a better chance of selecting the correct answer.

GLOSSARY

addend   a quantity that is added to another quantity. In the equation $x + 3 = 5$, x and 3 are addends.

additive inverse   the negative of a quantity

algebra   the representation of quantities and relationships using symbols

algebraic equation   an algebraic expression equal to a number or another algebraic expression, such as $x + 4 = -1$

algebraic expression   one or more terms, at least one of which contains a variable, and which may or may not contain an operation (such as addition or multiplication), but does not contain an equal sign

algebraic inequality   an algebraic expression not equal to a number or another algebraic expression, containing a $\neq$, $<$, $>$, $\leq$, or $\geq$, such as $x + 2 > 8$

arithmetic sequence   a sequence in which each term is found by adding a fixed value to the previous term. The sequence 3, 7, 11, 15, 19, . . . is an arithmetic sequence.

base   a number or variable that is used as a building block within an expression. In the term $3x$, x is the base. In the term 2^4, 2 is the base.

binomial   an expression that contains two terms, such as $2x + 1$

coefficient   the numerical multiplier, or factor, of an algebraic term. In the term $3x$, 3 is the coefficient.

composite number   a number that has at least one other positive factor besides itself and 1, such as 4 or 10

constant   a term, such as 3, that never changes value

coordinate pair   an x value and a y value, in parentheses separated by a comma, such as (4,2)

coordinate plane   a two-dimensional surface with an x-axis and a y-axis

cubic equation   an equation in which the highest degree is 3. The equation $y = x^3 + x$ is a cubic equation.

degree   The degree of a variable is its exponent. The degree of a polynomial is the highest degree of its terms. The degree of x^5 is 5, and the degree of $x^3 + x^2 + 9$ is 3.

distributive law   law stating that a term outside a set of parentheses that contains two terms should be multiplied by each term inside the parentheses: $a(b + c) = ab + ac$

dividend   the number being divided in a division problem (the numerator of a fraction). In the number sentence $6 \div 2 = 3$, 6 is the dividend.

divisor   the number by which the dividend is divided in a division problem (the denominator of a fraction). In the number sentence $6 \div 2 = 3$, 2 is the divisor.

domain   the set of all values that can be substituted for x in an equation or function

equation   two expressions separated by an equal sign, such as $3 + 6 = 9$

exponent   a constant or variable that tells you the number of times a base must be multiplied by itself. In the term $3x^2$, 2 is the exponent.

factor   If two or more whole numbers multiplied together yield a product, those numbers are factors of that product. Because $2 \times 4 = 8$, 2 and 4 are factors of 8.

factoring   breaking down a product into its factors

FOIL   an acronym that stands for First Outside Inside Last, which are the pairs of terms that must be multiplied in order to find the terms of the product of two binomials: $(a + b)(c + d) = ac + ad + bc + bd$

function   an equation in which every x value has no more than one y value

geometric sequence   a sequence in which the ratio between any term and the term that precedes it is always the same. The sequence 2, 4, 8, 16, 32, . . . is a geometric sequence.

imaginary number   a number whose square is less than zero, such as the square root of -9, which can be written as $3i$

inequality   two unequal expressions that are compared using the symbol $\neq$, $<$, $>$, $\leq$, or $\geq$

input/output table   a two-column table that shows examples of related values of x and y

integer   a whole number, the negative of a whole number, or zero. For example, 2 and –2 are integers.

like terms   two or more terms that have the same variable bases raised to the same exponents, but may have different coefficients, such as $3x^2$ and $10x^2$ or $7xy$ and $10xy$.

linear equation   an equation that can contain constants and variables, and the exponents of the variables are 1. For example, $y = 3x + 8$ is a linear equation.

mean   the quotient of the sum of all values in the set and the number of values in a set. For the set {1, 4, 5, 6}, the mean is $(1 + 4 + 5 + 6) \div 4 = 4$.

median   the value that is in the middle of a data set after the set is put in order from least to greatest. If there is an even number of terms, the median is the mean of the two middle terms.

monomial   an expression that contains only one term, such as $3x^2$

ordered pair   an x value and a y value, in parentheses separated by a comma, such as (4,2)

parallel lines   lines that have the same slope. Parallel lines never intersect.

percent   a number out of one hundred. For example, 36% is 36 out of 100.

perpendicular lines   lines that meet or cross at right angles

polynomial   an expression that is one term or the sum of two or more terms, such as $x^2 + 2x + 1$, each with whole-numbered exponents

prime factorization   the writing of a number as a multiplication expression made up of only prime numbers, the product of which is the original number

prime number   a number whose only positive factors are 1 and itself, such as 3 or 7

probability   the likelihood that an event or events will occur, usually given as a fraction in which the numerator is the number of possibilities that allow for the event to occur and the denominator is the total number of possibilities

product   the result of multiplication. In the number sentence $2 \times 4 = 8$, 8 is the product.

proportion   an equation that shows two equal ratios, such as $\frac{16}{12} = \frac{4}{3}$

Pythagorean theorem   an equation that describes the relationship between the sides of a right triangle, showing that the sum of the squares of the bases (legs) of the triangle is equal to the square of the hypotenuse of the triangle: $a^2 + b^2 = c^2$

quadratic equation   an equation in which the highest degree is 2. For example, $y = x^2 + 1$ is a quadratic equation.

quotient   the result of division. In the number sentence $6 \div 2 = 3$, 3 is the quotient.

radical   a root of a quantity

radicand   the quantity under a radical symbol. In $\sqrt[3]{8}$, 8 is the radicand.

range   the set of all y values that can be generated from x values in an equation or function; also, the difference between the greatest and least values of a data set

ratio   a relationship between two or more quantities, such as 3:2

root   a value of x in a function for which $f(x)$ is 0

sequence   a set of numbers in which each number is generated according to a rule

slope   the change in the y values between two points on a line divided by the change in the x values of those points

slope-intercept form $y = mx + b$, where m is the slope of the line and b is the y-intercept

system of equations   a group of two or more equations for which the common variables in each equation have the same values

term   a variable, constant, or product of both, with or without an exponent, that is usually separated from another term by addition, subtraction, or an equal sign, such as $2x$ or 5 in the expression $(2x + 5)$

trinomial   an expression that contains three terms, such as $6x^2 + 11x + 4$

unknown   a quantity whose value is not given, usually represented by a letter

unlike terms   two or more terms that have different variable bases, or two or more terms with identical variables raised to different exponents, such as $3x^2$ and $4x^4$

variable   a symbol, such as x, that takes the place of a number

vertical line test   the drawing of a vertical line through the graph of an equation to determine if the equation is a function. If a vertical line can be drawn anywhere through the graph of an equation, such that it crosses the graph more than once, then the equation is not a function.

x-axis   the horizontal line on a coordinate plane along which $y = 0$

y-axis   the vertical line on a coordinate plane along which $x = 0$

y-intercept   the y value of the point where a line crosses the y-axis

NOTES

NOTES

NOTES

NOTES

NOTES